gas chromatography
in the analysis
of steroid hormones

gas chromatography in the analysis of steroid hormones

Herbert H. Wotiz
Professor of Biochemistry

and

Stanley J. Clark
Research Associate in Biochemistry
Boston University, School of Medicine
Boston, Massachusetts

℗ Springer Science+Business Media, LLC 1966

ISBN 978-1-4899-6618-6   ISBN 978-1-4899-6633-9 (eBook)
DOI 10.1007/978-1-4899-6633-9

Library of Congress Catalog Card Number 65-11333

© 1966 Springer Science+Business Media New York
Originally published by Plenum Press in 1966.
Softcover reprint of the hardcover 1st edition 1966

To M & M, our lives' sustenance

Preface

Gas chromatography is a powerful tool for the assay of steroidal hormones and its use is spreading rapidly. New analytical methods appear in the journals every month, as do revisions of methods published only recently. The speed, sensitivity, and resolving power of gas chromatography have allowed the development of simpler, faster, and more sensitive assay methods that are attractive, not only to the research worker, but also to the clinician requiring rapid and accurate measurement of hormone output.

However, before the newcomer can derive the maximum benefit from the technique, it is necessary to collect and to evaluate the information scattered through the recent literature. It is hoped that this book may spare the reader much of the labor.

During the past five years, the authors have been deeply involved in the practice of gas chromatography both in the development of instrumentation and in the application of the technique to steroid analysis. Early interest in this work by other investigators prompted the establishment of a course in the gas chromatography of steroids, given annually for the past three years at Boston University. The subsequent successful establishment of gas chromatographic procedures by many of the participants encouraged the authors to revise and expand the material presented at the course. This book is the result.

Two classes of methods are presented: those with which the authors have extensive personal experience and those for which published reports contain sufficient detail so that a fair assessment of the merit of the method is possible. The book is not a general text on gas chromatography and only those aspects of theory and practice relevant to the assay of steroids are considered. The individual analytical methods, however, are treated in great detail so that the newcomer either to endocrinology or to gas chromatography may readily set up procedures for routine use.

April 1966

Herbert H. Wotiz
and
Stanley J. Clark

Acknowledgments

No book is ever written without the material assistance and sound advice of one's colleagues and fellow workers. This book is no exception.

The authors wish to thank Professor F. M. Sinex for his encouragement and enthusiastic support in this venture and Mrs. Joyce Smith for her valiant efforts at typing and retyping the manuscript. To our co-workers R. Aldin, P. Brecher, G. Charransol, G. Rudnicki, I. Sarda, F. Stendler, A. Stranieri, E. Wilson, B. P. Ying, Drs. S. C. Chattoraj, R. Guerra-Garcia, H. E. Carr, Jr., I. S. Hartman, and H. F. Martin we extend our gratitude for their contributions in the laboratory and production of results which have added so much to the text.

Thanks are due a number of investigators whose generosity and cooperation made available to us manuscripts prior to their publication so that their contents might be incorporated here. Among these contributors were E. Bailey (London), A. Brownie (Buffalo), M. Lipsett (Washington), I. Menini (Edinburgh), R. Rosenfeld (New York), D. Sandberg (Miami), I. Somerville (London), M. Sparagana (Boston), B. S. Thomas (London), and H. J. VanderMolen (Utrecht).

We are especially grateful to those of our colleagues who not only provided information but actually wrote certain sections of the book which fell within their special fields of interest. Chapter 13 was written entirely by Dr. I. S. Hartman (Boston), who in turn made use of information generously supplied by Drs. B. Knights (Glasgow) and J. Chamberlain (Boston). Dr. S. C. Chattoraj (Boston) contributed the bulk of the text on Chapter 17 (estrogens) and Dr. R. Rosenfeld (New York) provided certain sections of Chapter 15 (pregnanetriol) and Chapter 16 (corticosteroids). Mrs. E. Smakula has provided all the spectroscopic data from our laboratory presented here as evidence for molecular integrity and methodological specificity.

The authors also wish to acknowledge the cooperation of the F & M Scientific Corporation, Avondale, Pennsylvania; Perkin-Elmer Corporation, Norwalk, Connecticut; and Jarrell-Ash Company, Waltham, Massachusetts, who have made available material reproduced in the text. Further thanks are expressed to these companies as well as Wilkens Instrument & Research, Research Specialties Company, Bar-

ber-Colman Company, and Packard Instruments, whose active partici-
pation in the GLC courses at Boston University Medical Center has
contributed to their success.

Last, but not least, grateful acknowledgment is made of the
assistance rendered by the National Institutes of Health for the
awarding of a Research Career Development award (GM-15,369) and
a Research Grant (CA-03135), without which the information in this
book could not have been compiled.

Contents

Abbreviations

GLC = gas–liquid chromatography
HETP = height equivalent to a theoretical plate
n = number of theoretical plates
t_r = retention time
r = relative retention
R = resolution
TLC = thin-layer chromatography
psi = pounds per square inch
PTFE = polytetrafluoroethylene
OD = outside diameter
ID = inside diameter
TMSi = trimethylsilyl ether
QF-1 = fluoroalkyl silicone polymer
SE-30 = dimethyl silicone elastomer
XE-60 (CNSi) = silicone nitrile polymer
EGIP = ethylene glycol isophthalate polyester
NGS = neopentylglycol succinate polyester
Hi-Eff 8B (CDMS) = cyclohexane dimethanol succinate polyester
E_1 = estrone
E_1A = estrone acetate
E_2 = estradiol
E_2A = estradiol diacetate
E_3 = estriol
E_3A = estriol triacetate

Chapter 1

Introduction

The stimulus that led to the development of liquid–liquid partition chromatography was the need for better separatory techniques in biochemistry. Similarly, gas–liquid partition chromatography had its origin in biochemistry, although its early development was largely in the fields of organic and petroleum chemistry. Compared with the time lag between the original discovery of chromatography and its widespread use or, for that matter, between the suggestion that gas–liquid partition chromatography was possible and the experimental demonstration that confirmed this suggestion, the exploitation of gas chromatography has been remarkably rapid. The spectacular successes achieved in separating hitherto intractable mixtures of organic compounds, together with the early development of commercial equipment, were undoubtedly major factors contributing to this rapid growth. However, several years elapsed before instrumentation of sufficient refinement was developed to deal with large organic molecules of biochemical interest. The pattern of rapid development then repeated itself in the field of interest, namely, the gas chromatography of steroids, so that in a period of only five years enough has been done to establish the great utility of the technique. During the past three years especially, many quantitative methods for the assay of steroidal hormones and their metabolites have been reported.

The methods described in this book may be divided into three broad categories. In the first are those methods designed to deal with relatively high concentrations of the compound sought where the primary requirement is speed of analysis. Separation of the original sample into gross fractions followed by preparation of derivatives is the only preliminary to gas chromatography, reliance being placed entirely upon the gas chromatographic column to achieve the necessary separation. The elapsed time for methods in this class is usually two to three hours so that the results obtained are clinically useful. The second category consists of rather more complex methods which allow measurement of a number of compounds simultaneously. Generally some preliminary separation is necessary before

1

gas chromatography, usually employing thin-layer chromatography. Somewhat greater sensitivity and specificity is achieved at the expense of speed, but the analysis can generally be completed in twenty-four hours and often in less time. Finally, in the third category, are those methods that place a severe demand even upon the highly sensitive electron capture gas chromatography system. In order to detect a compound present in extremely small amounts, it is necessary to subject the sample to thorough purification. Thus procedures in this group are lengthy, but achieve sensitivities previously unattainable.

There exists a strong tendency to become overenthusiastic about a new technique and to attempt to apply it even when it is obvious that for one reason or another, other methods are better suited to the problem in hand. At present, for example, not all the corticosteroids can be dealt with by gas chromatography. Although the technique is very well suited to the assay of many steroids, its application is probably not universal.

Gas chromatography, therefore, must not be regarded as a panacea and the reader must use his judgment in deciding whether the methods presented here are applicable to his problem and must then make use of the tool best suited to do the job.

Chapter 2

Definition of Chromatography

Chromatography is a method of separation based on differential migration. The separation of mixtures into individual components depends upon the differential sorption of the components by an active stationary phase. Migration through the stationary phase is produced by a driving force—the flow of a usually nonselective, mobile fluid phase. In gas–liquid chromatography one is concerned with that major subdivision of chromatography in which the mobile phase is an inert gas.

GAS CHROMATOGRAPHY

The heart of the gas chromatographic system is the column, which is a long tube packed with a sorbent material. The column packing is permeable and an inert gas, the carrier gas, passes continuously through it. In all the applications to be considered here, the column packing consists of an inert support of large surface area coated with a film of involatile liquid. Separation is achieved through differential partition of the sample between the involatile liquid, or stationary phase, and the carrier gas, or mobile phase. Thus, the technique is known as gas–liquid partition chromatography, usually shortened to gas–liquid chromatography or gas chromatography.

The sample consists of either a vaporizable liquid or solid which is introduced as nearly instantaneously as possible into the carrier gas stream at the head of the column. The components of the mixture are detected as they emerge by a detector at the end of the column which measures either a change in property of the carrier gas or, better, some property of the component itself. Electrical signals from the detector are recorded on a strip-chart recorder.

THE COLUMN

The sample vapor which is introduced into the column is largely sorbed by the stationary phase, but tends to set up an equilibrium with

3

the mobile phase by reason of its vapor pressure over the stationary phase. That is to say, it is partitioned between the two phases. The material in the mobile phase moves along the column, where it redissolves in the stationary phase and again tends to come to equilibrium. At the same time the material already dissolved re-enters the gas phase so as to restore equilibrium. The process repeats itself continuously so that the sample moves down the column as a more or less compact band or zone.

The speed of movement of any component of a mixture along the column is governed by its thermodynamic behavior: that is to say, it depends upon the extent of partition between stationary and mobile phases. How well the column achieves a given separation depends upon the extent to which the zone spreads during its passage through the column. Both aspects of column performance must be considered in the development of gas chromatographic theory.

Derivations of all the relationships given in the following sections can be found in the standard textbooks on gas chromatography. It would be pointless to reproduce them here since we are concerned primarily with the practical consequences of the effect of experimental conditions on performance. It should be realized that development of column theory is still in progress and that what is presented here is much simplified and is intended only to give a qualitative picture of what happens in a chromatographic column.

Chapter 3

Gas Chromatographic Columns

COLUMN THEORY
Retention Volume

An idealized chromatogram of a pure compound is shown in Figure 3-1. Injection is made at point A, and any material in the sample, such as air, that is not soluble in the stationary phase is swept through the column at the same speed as the carrier gas, emerging at point B. A compound soluble in the stationary phase will travel more slowly, emerging at C. The time from the point of injection to the peak maximum is the retention time t_r (AC in Figure 3-1), and the volume of gas passed through the column during this time is the retention volume V_r. The quantities are related as follows:

$$t_r = \frac{V_r}{f_{av}}$$

where f_{av} is the average flow rate of gas through the column. In most applications, flow rate is held constant so that measuring retention time on the recorder chart is equivalent to measuring retention volume. The chromatogram illustrated is an example of a differential chromatogram. This form of display is virtually universal and will be treated here to the exclusion of all others.

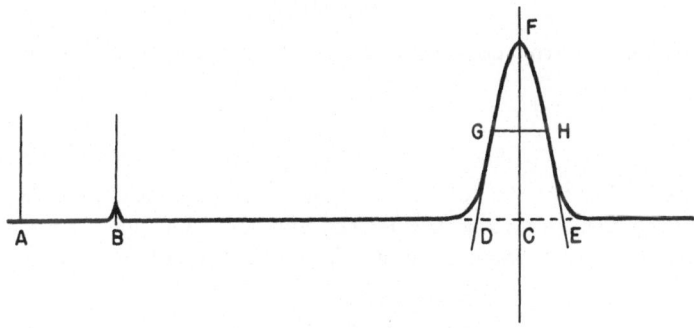

Figure 3-1. Idealized chromatogram.

5

The trace drawn by the recorder in the absence of sample is known as the b a s e l i n e . Ideally, it should be parallel to or coincident with the zero line on the recorder chart. The trace drawn in response to a sample component passing through the detector is known as a p e a k . As the band of sample vapor moves along the column, it spreads out and the width of the band relative to the retention time is a measure of column efficiency. The peak should be approximately Gaussian in shape, which might be expected since the shape reflects the randomness of the processes causing peak spreading. The p e a k w i d t h W is found by drawing tangents to the points of inflection of the curve and measuring the distance (DE, Figure 3-1) between the intersections of these tangents with the baseline. It is usually more convenient to measure the p e a k w i d t h at h a l f - h e i g h t , $W_{1/2}$, (GH in Figure 3-1). The distance CF measured perpendicular to the baseline is the p e a k h e i g h t , and the area enclosed between the peak and baseline is the p e a k a r e a .

The distance AB in Figure 3-1 represents the retention volume of a nonabsorbed sample. It is the sum of the volume of the column not occupied by packing and the volumes of injection system, detector, and any connecting tubing. These latter are assumed to be negligible in the theoretical discussion although this is not always true in practice. This volume V_g is variously known as d e a d v o l u m e or g a s h o l d - u p v o l u m e . It is advantageous to define an a d j u s t e d r e t e n t i o n v o l u m e V_r' such that

$$V_r' = V_r - V_g$$

which is equal to BC in Figure 3-1.

The carrier gas is compressible so its velocity is not uniform, but increases as it passes through the column. The adjusted retention volume may be corrected for the nonuniformity of flow to give the n e t r e t e n t i o n v o l u m e V_n, where

$$V_n = j V_r'$$

The pressure drop correction factor j is found from the relation:

$$j = \frac{3}{2} \cdot \frac{(p_i/p_o)^2 - 1}{(p_i/p_o)^3 - 1}$$

where p_i and p_o are the column inlet and outlet pressures, respectively.

All the parameters so far defined are dependent upon the weight of stationary phase present. The s p e c i f i c r e t e n t i o n v o l u m e V_s is defined as

$$V_s = \frac{V_n \cdot 273}{W_l T}$$

where W_l is the weight of liquid phase in the column and T is temperature (°K). This quantity is the retention volume per gram of stationary phase (reduced to 0°C) and is independent of the column.

It is often advantageous to express retention volumes relative to a standard compound chromatographed under the same conditions. The relative retention r is given by

$$r = \frac{V_{s_1}}{V_{s_2}}$$

where V_{s_2} is the specific retention volume of the standard. It should be noted that relative retentions must be calculated from adjusted retention volumes or related parameters.

Partition Coefficients

The speed of movement of any compound through the column is dependent upon the partition coefficient K

$$K = \frac{\text{Weight solute per ml stationary phase}}{\text{Weight solute per ml mobile phase}}$$

Taking into account the differences in volume between gas and liquid phases, it can be shown that

$$Z = \frac{V_g}{V_g + KV_l}$$

where Z is the fraction of solute in the gas phase, V_g is the gas hold up volume of column, and V_l is the volume of stationary phase; and also that

$$V_r = KV_l + V_g = V_r' + V_g$$

Thus, the retention volume is related to the partition coefficient for a particular compound on a particular column.

Effect of Temperature on Retention Volume

The partition coefficient is related to temperature as follows:

$$K = \frac{RT\rho}{\gamma p^\circ M}$$

where ρ is the density of stationary phase, M is the molecular weight of stationary phase, γ is the activity coefficient, and p° is the vapor pressure of pure solute.

Taking into account the known effect of temperature on vapor pressure and activity coefficient, we can write

$$\log V_s = \frac{\Delta H}{2.3\ RT} + k$$

where ΔH is the partial molar heat of evaporation of solute from solution and k is constant. Variation of ΔH with temperature is usually small over the short range of interest for gas chromatography and thus plots of log retention volume against the reciprocal of absolute temperature are linear.

Although the assumption that ΔH is constant holds reasonably well for many solute—solvent pairs, it is often found that some curvature occurs in log V_s plots. In these cases, an Antoine equation of the form

$$\log V_s = A + \frac{B}{t + C}$$

where t is temperature (°C) and A,B, and C are experimentally determined constants will usually result in a linear plot.

Relative retention is much less affected by temperature since the difference between two partial molar heats of evaporation enters into the equation and this value is always much less than ΔH for either of the solutes.

Plate Theory

Although gas chromatography is a continuous process, it is convenient to introduce the concept of the theoretical plate which derives from an analogous but discontinuous process. The column may be regarded as divided into a number of isolated "plates" joined only by the flow of carrier gas. The theoretical plate may then be defined as the smallest length of column in which the partition process can come to equilibrium.

The analogy of countercurrent distribution may be used to derive expressions for peak shape and related quantities [1, 2].

Consider a series of separatory funnels into each of which is placed a volume V of solvent. This solvent is the stationary phase S. To the first funnel is added an equal volume V of a second solvent (the mobile phase M), immiscible with the first, containing W grams of a solute. For the sake of simplicity it is assumed that the partition coefficient K is unity. That is, at equilibrium, the solute will be distributed equally between the two phases. The mixture in the first funnel is equilibrated and the mobile phase transferred to the second funnel. Fresh mobile phase (volume V) is added to the first funnel. The process is repeated over all the funnels, each portion of mobile phase being successively transferred to the next funnel and

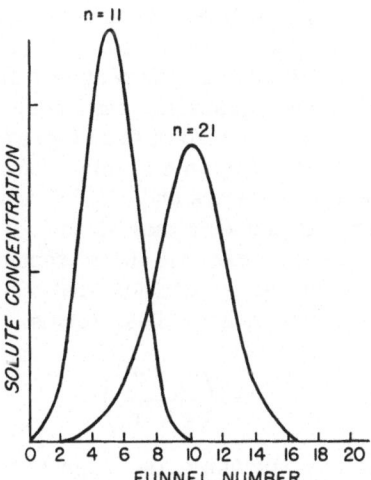

Figure 3-2. Solute distribution after 11 and 21 extractions. [A.I.M. Keulemans, "Gas Chromatography", Rheinhold, 1959].

successive fresh portions added to the first. The diagrams in Figure 3-2 show the distribution of the solute in a small number of funnels after different numbers of extractions. The weight of material in any funnel is given by the terms of the expansion of the binomial:

$$\left(\frac{1}{K} + \frac{K}{K+1}\right)^n$$

where n is the number of funnels and K is the partition coefficient.

When n is large, as in gas chromatography, the binomial distribution can be transformed into a Gaussian distribution, giving an equation for the curve that may be written

$$Q_n = \frac{1}{\sqrt{2\pi n}} \left(\frac{V}{V_g + KV_l}\right)^n e^{n \cdot V/V_h}$$

where Q_n is the quantity of material in plate n, V is the volume of gas passed through plate n, and V_h is the volume of a single plate.

The curve has a maximum at $V = V_r = nV_h$ and the peak width $W = 4V_h \sqrt{n}$ and thus $n = 16 (V_r/W)^2$.

It may be preferable from a practical point of view to measure the width at half height $W_{1/2}$:

$$n = 5.545 \left(\frac{V_r}{W_{1/2}}\right)^2$$

The larger the number of theoretical plates, the greater the separatory power of the column will be. At the same time, the peak width is proportional to the plate volume so that, for a given number of plates, the smaller V_h, the sharper the peak and the better the column will be. V_h is not directly measurable but a comparable quantity, the height equivalent to a theoretical plate (HETP), is given by L/n, where L is the length of column in mm. Thus, specification both of the number of theoretical plates and of HETP gives a better picture of column performance than does either alone.

So far, only a single peak has been considered, but since the analyst's interest is in the separation that can be achieved, it is useful to characterize columns in these terms. The resolution R is given by

$$R = 2\left(\frac{V_{r_1} - V_{r_2}}{W_1 + W_2}\right)$$

Thus for a resolution of unity, the tangents to the adjacent sides of the peaks intersect the baseline at the same point, the overlap of the peaks is small, and separation is virtually complete. Resolutions of less than unity signify incomplete separation.

If the relative retention of the two peaks is known, the number of plates required to achieve any given resolution may be calculated from:

$$n = \left[\frac{2R(r + 1)}{r - 1}\right]^2$$

Peak Distortion

Certain assumptions are implicit in the simplified discussion given above, which, if not taken into account, result in considerable deviation of peak shape from that derived theoretically.

In the first place, the derivation of the Gaussian distribution curve is based on the assumption that the sample initially occupies only the first plate. In fact, the sample normally occupies a number of plates so that the chromatogram starts from a number of different points along the column. Thus not only is the peak broadened, but the retention volume will differ slightly from the theoretical value. Obviously, the number of plates occupied will depend on the amount of sample, and thus column efficiency and retention volume are somewhat dependent upon sample size.

Secondly, the assumption that the vapor pressure of the solute and the solvent, and hence the partition coefficient, is independent of concentration is true only in special circumstances. The fundamental relationship for the vapor pressure of a volatile solute over its solution is

$$p_1 = \gamma_1 n_1 p_1^\circ$$

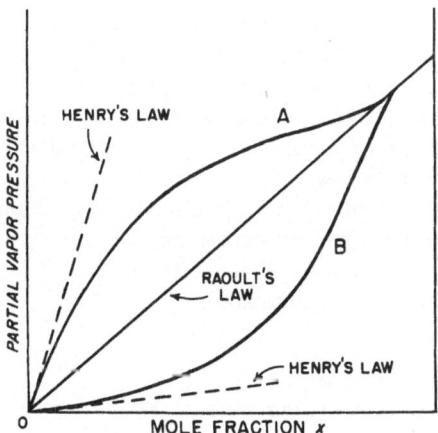

Figure 3-3. Deviations of activity coefficients from ideality.

where p_1 is the vapor pressure of the component, n_1 is its mole fraction in the solution, p_1° is the vapor pressure of the pure solute under the same conditions, and γ_1 is its activity coefficient.

When γ_1 is constant the relationship is known as Henry's law and, when γ_1 is both constant and has a value of unity, as Raoult's law. Figure 3-3 illustrates two practical cases: A having positive deviation from Raoult's law ($\gamma > 1$) and B having negative deviation ($\gamma < 1$). At very low concentrations of solute, both curves approach linearity and in these regions Henry's law holds reasonably well.

In gas chromatography, attempts are always made to operate at very low concentrations of solute, but even though this condition may hold true for the bulk of the column, it often happens that a short section of the front of the column is overloaded when a sample is injected. Thus, linear conditions are only attained when the sample has been diluted by moving some distance along the column. For positive deviations from linearity, the partition coefficient increases as the concentration increases and the peak maximum is delayed with respect to the skirts of the peak where the concentration is lower. The resulting peak has a trailing edge sharper than its leading edge

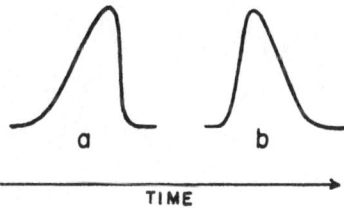

Figure 3-4. Peak skewing arising from nonlinear conditions.

(Figure 3-4a). For negative deviations, conversely, the regions of higher concentration move faster than those of lower concentration with the result that the peak "tails" (Figure 3-4b).

The support used to carry the stationary phase is rarely completely inert and its adsorptive properties have the same effect as a negative deviation from Henry's law.

Nonlinearity problems are particularly important in steroid gas chromatography since, for practical reasons, only small amounts of stationary phases are used. Thus, effects both of overloading and of support adsorptivity are accentuated.

Measurement of Retention Time

The practice of measuring retention time from points other than the peak maximum is, unfortunately, becoming all too common [2a]. The practice apparently originated as an attempt to deal with chromatograms obtained from nonlinear columns where badly skewed peaks were recorded. Most often, peaks with exaggerated tails are produced and retention times are then measured from the point at which the leading edge of the peak starts to rise from the baseline. The procedure must be criticized on several counts: First, at the present stage of development of steroid gas chromatography, nonlinearity can be almost entirely eliminated by proper attention to experimental detail. The only prerequisite is that the investigator be competent. Second, the procedure has no theoretical justification. Third, there is no guarantee that measurement from the peak front is any more precise than measurement from the peak maximum. In the all too common case where peaks overlap, the position of the minimum between the peaks is dependent upon the relative amounts of the two components (Figure 3-5) and measurements to this point have no real meaning.

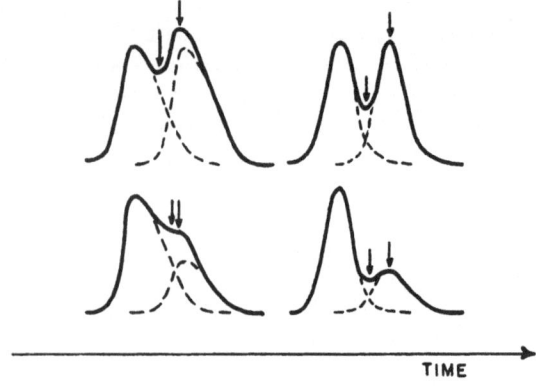

Figure 3-5. Illustration of errors in measurement of retention time.

Fourth, if, despite the best efforts of the experimenter, nonlinear peaks persist, retention times should be determined by the well-established method of extrapolation to zero concentration. This approach is admittedly somewhat time-consuming, but does lead to valid results.

Selectivity of Stationary Phase

Separation of a mixture is only achieved if the components of the mixture have different partition coefficients. In other words, the quantity $1/p^{\circ}\gamma$ must have a different value for each component (see page 7). For compounds having different vapor pressures it is only necessary to ensure that the stationary phase used does not change γ in such a way as to diminish the separation. Separations of chemically similar compounds differing only in molecular weight are examples of this type. On the other hand, separation of two compounds having similar vapor pressures requires that the activity coefficients differ. If the compounds differ in chemical type, it is usually possible to select a stationary phase in which the solute–solvent interaction is stronger for one compound. However, isomers having nearly identical vapor pressures may be very difficult to separate.

COLUMN PERFORMANCE

In the development of the plate theory, the assumptions are made that each plate is discrete and that equilibrium is attained in each plate. In fact, gas chromatography is a continuous process in which the sample vapor is free to diffuse in all directions in both gas and liquid phases. Furthermore, the partition process never comes to complete equilibrium as the sample travels along the column. Any theory which attempts to describe the effect of experimental variables upon column performance must take into account both diffusion and non-equilibrium processes.

The first general approach to the problem was developed by van Deemter et al. [3]. There can be no doubt that, despite certain deficiencies, the original theory and later modifications of it account for the major processes contributing to band spreading. Experimental tests have not only provided broad confirmation of the theory, but have often shown the way to considerable improvement in column performance.

The original van Deemter equation has the form

$$H = A + B/\bar{u} + C\bar{u}$$

where H is height equivalent to a theoretical plate and $\bar{u}$ is average linear gas velocity in the column. A plot of H against $\bar{u}$ is shown in Figure 3-6, indicating the contribution of each term of the equation.

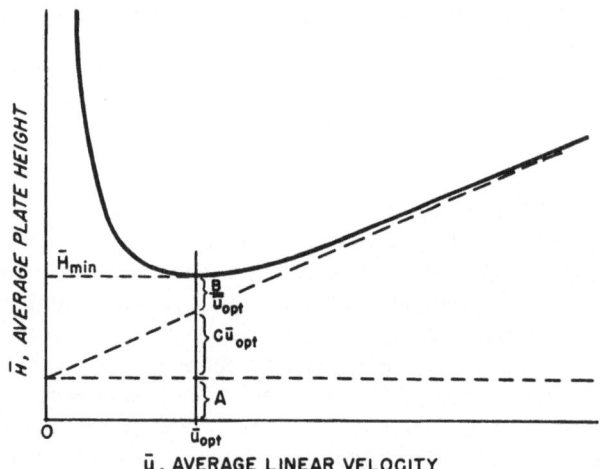

Figure 3-6. Simplified van Deemter equation: Relation between plate height and gas velocity.
[A. I. M. Keulemans, "Gas Chromatography", Rheinhold, 1959].

It is of practical importance to determine optimum conditions for a given column by measuring H as a function of $\bar{u}$ experimentally.

The first term represents the contribution of unequal path lengths through the column packing to band spreading, the second of longitudinal diffusion in the gas phase, and the third of nonequilibrium in the partition process. In later developments, the third term is split into two parts and an extended equation may be written:

$$H = A + B/\bar{u} + C_l \bar{u} + C_g \bar{u}$$

Spreading Due to Differences in Path Length

The first term, A, accounts for the fact that gas passages through the column packing are not all identical. Thus, the time taken for a solute molecule to pass through a given section of column packing will depend upon the path taken by the molecule and the solute band will spread. The coefficient A is given by

$$A = 2\lambda d_p$$

where λ is a factor reflecting the randomness of the packing and d_p is the average particle diameter. The magnitude of λ has been the subject of much study and values ranging from about 0.5 to 8 have been reported. Whatever the value, the effect of randomness of packing will be obscured if the column is so badly packed that channeling occurs. Thus, the first requirement for good column performance is that the packing be loaded into the column as uniformly as possible.

It would be expected from the equation that A would decrease as the particle size became smaller and more uniform and this appears to be generally true. The main disadvantage of using very small particles is that high inlet pressures are then required to maintain gas flow through the column. Instrument limitations and the difficulty of introducing samples against high inlet pressures limit the pressures that can be employed.

Diffusional Spreading

Differences in retention time arise from differences in partition coefficient and are determined mainly by the length of time spent by a solute in the liquid phase. All solutes spend about the same time in the mobile phase and this time is determined by the column dead volume and the average gas velocity. During its time in the mobile phase the solute is free to diffuse through the mobile phase, with the result that the solute band broadens. This diffusional spreading is represented by the second term of the van Deemter equation:

$$B/\bar{u} = \frac{2\gamma D_g}{\bar{u}}$$

where D_g is the diffusion coefficient of the solute vapor in the carrier gas and γ is a factor representing the effect of the packing material on gaseous diffusion.

The gas velocity (or flow rate) is most conveniently measured at the outlet of the column but the value $\bar{u}$ in the van Deemter equation is the measured value corrected for the pressure drop (see page 6), i.e., $\bar{u} = u_o j$.

The second term of the equation, $2\gamma D_g/\bar{u}$ contains D_g, the diffusivity of the solute in the carrier gas which itself is pressure-dependent and which varies in the same manner as the local gas velocity. Thus at any particular point in the column, $D_g/\bar{u}$ is independent of pressure. For this reason the second term of the equation is more conveniently written $2\gamma D_g^o/u_o$, where the values of D_g^o and u_o are those corresponding to the pressure at the outlet of the column. When $\bar{u} \simeq u_o$, which is the case when the pressure drop across the column is small, the plot of H against $\bar{u}$ is meaningful, but when this condition does not apply there is no simple way of illustrating the relationship.

Values between 0.5 and 1 have been reported for γ. If the true value is, in fact, unity, then the column packing should have no effect on gaseous diffusion. More experimental work is required to clarify this point.

From the dependence of B_o upon D_g^o it would be expected that a higher molecular-weight carrier gas would yield higher column efficiencies since D_g^o decreases with increasing molecular weight. However, the extended rate equation contains the $C_g^o u_o$ term which contains the ratio u_o/D_g^o (see below). At optimum gas velocity (H is at minimum) the effect of the two terms will cancel insofar as they are dependent upon D_g^o and thus no improvement in efficiency will result from changing carrier gas. It should be noted, however, that the optimum gas velocity is dependent upon D_g^o and thus will have different values for different carrier gases. At gas velocities other than optimum, however, the situation is different: At low velocities $(B_o/u_o) > C_g^o u_o$, while at high velocities $C_g^o u_o > (B_o/u_o)$. Thus, carrier gases of high molecular weight will give lower values of H at low velocities, and low molecular-weight gases will give the best performance at high velocities.

The effect of operating pressure on performance again depends on the relative values of B_o/u_o and $C_g^o u_o$. Thus at optimum velocity, the value of H will be independent of pressure; at low velocities high pressures will improve performance and at high velocities low pressures are desirable.

It should be noted that the increase in operating pressure rather than change in the ratio p_i/p_o explains the results reported by Scott [4, 5] and that, in general, pressure drop across the column has little effect on performance.

Resistance to Mass Transfer

In any plate of the column, transfer of solute into and out of the stationary phase occurs and at the same time solute is carried along the column in the flow of carrier gas. Solution and evaporation of solute molecules are not instantaneous processes and during the time that solute molecules reside in the stationary phase they are free to diffuse through the liquid layer. The contribution to band spreading of the mass transfer process is given by

$$C_l \bar{u} = \frac{2k'd_f^2 \bar{u}}{3(1 + k')^2 D_l}$$

where d_f is the effective thickness of the stationary phase layer, D_l is the diffusion coefficient of the solute in the stationary phase, and $k' = KV_l/V_g$.

Evaluation of this term is difficult because the quantity d_f is ambiguous. Ideally, the stationary phase should be distributed over the support as a uniform, continuous film and in special cases, for example, in capillary columns or glass bead columns where the support surface is well defined, this is so. For thin film distribution, it can be shown that

$$C_l = \frac{(k')^3 d_p^2}{24K^2 D_l (1 + k')^2}$$

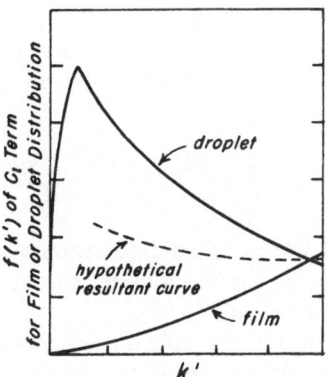

Figure 3-7. Effect of type of stationary phase distribution upon C_l. [H. Purnell, "Gas Chromatography", John Wiley, 1962].

All the quantities in this expression can be evaluated experimentally and results with capillary columns confirm the relationship.

In normal packed columns, employing a diatomaceous earth support, the situation is more complex. It seems likely that when very small amounts of stationary phase are applied, the stationary phase is distributed as droplets. As the amount of stationary phase is increased, the pores of the support fill up and eventually film distribution will be achieved.   In practice, columns are almost always operated in the transition region where both types of distribution occur. Figure 3-7 [6] shows the effect upon plate height of the C_l term for each type of distribution as k' increases.   All other variables are held constant and k' may be measured as the ratio of stationary phase to support. The dotted curve indicates what may happen for a mixed stationary phase distribution.

Experimental results offer qualitative confirmation in that a minimum occurs at about 18% (w/w) of stationary phase. Column efficiencies decrease as the amount of stationary phase increases or decreases.   Fortunately, the change in efficiency over the range 1–20% of stationary phase is not large, but it seems likely that a substantial gain in efficiency would result at low percentages of stationary phase if film distribution could be achieved.

Radial Diffusion in the Gas Phase

In the development of the theory for capillary columns, it was found that a term in the equation was required to account for diffusion in the gas phase in the radial direction, i.e., at right angles to the direction of flow.   It would seem likely that such diffusion will be even more important in packed columns because of the irregularity of the flow paths through the packing. At the same time, lack of knowledge of velocity profiles through packed beds makes it difficult to determine the exact form of the term. Purnell [7], by analogy with capillary systems, gives

$$C_g^o = \left[\frac{1 + 6k' + 11(k')^2}{24(1 + k')^2} \right] \frac{d_p^2 \, \chi}{D_g^o}$$

where χ is a factor which takes into account the irregularity of the packed bed. Note that, as for the B term, values of u and D_g are those obtaining at the column outlet.

As for the C_l term, column efficiency should increase as particle size decreases. The function of k' increases with k' up to a value of 10 and then remains constant so that for good solvents (i.e., for late-emerging components) the solvent–support ratio has no effect on efficiency. As discussed earlier, the B and C_g terms must be considered together to determine the effect of carrier gas and velocity upon efficiency.

REFERENCES

1. A. J. P. Martin and R. L. M. Synge, Biochem. J. 35:1358 (1941).
2. A. T. James and A. J. P. Martin, Biochem. J. 50:679 (1952).
2a. A. A. Patti and A. A. Stein, Steroid Analysis by Gas Liquid Chromatography (Charles C Thomas, Springfield, Illinois, 1964).
3. J. J. van Deemter, F. J. Zuiderweg, and A. Klinkenberg, Chem. Eng. Sci. 5:271 (1956).
4. J. D. Cheshire and R. P. W. Scott, J. Inst. Petrol. 44:74 (1958).
5. R. P. W. Scott, Gas Chromatography, edited by D. H. Desty (Butterworths, London, 1958), p. 189.
6. J. H. Purnell, Gas Chromatography (John Wiley & Sons, Inc., New York, 1962), p. 149.
7. J. H. Purnell, Gas Chromatography (John Wiley & Sons, Inc., New York, 1962), p. 136.

Chapter 4

Instrumentation

To achieve an analysis, the sample must be presented to the chromatographic column in a suitable form. The column operating conditions must be properly maintained and some means must be provided both to determine what separation has been achieved and to measure the separated components of the sample. Instrumentation to serve these functions is manufactured by about thirty domestic companies, of which about a dozen can be considered major suppliers; probably, between fifty and one-hundred models and variants are offered by these companies. Obviously, the newcomer to gas chromatography is likely to be bewildered by the variety of equipment offered to him. It would be inappropriate, here, to recommend the products of a particular manufacturer, but a discussion of the instrumental requirements for steroid gas chromatography together with some indication of the features critical to success can help the novice to select equipment best suited to his needs.

A block diagram of a complete chromatographic system is shown in Figure 4-1. The only item not essential to the chromatographic process is the block labeled "ancillary methods," but, since gas

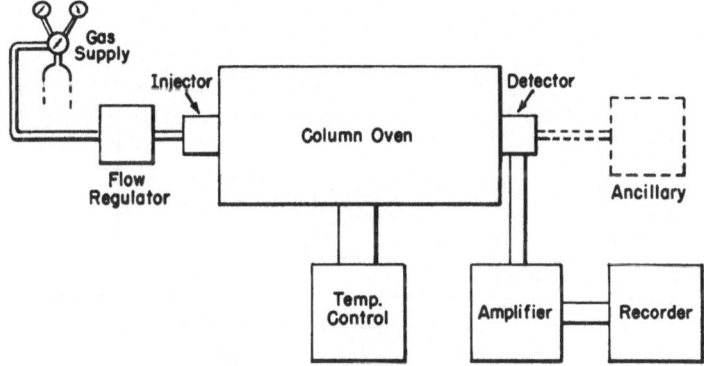

Figure 4-1. Gas chromatographic system.

19

chromatography is a rather poor qualitative tool, it is appropriate here to consider techniques for the unequivocal identification of compounds separated by gas chromatography.

GAS CONTROLS

Carrier gases are normally supplied in high-pressure cylinders and some means of reducing and controlling the inlet gas pressure from the cylinder is required. Regulators suitable for the purpose are available from most gas suppliers. A maximum delivery pressure of 100 psi is adequate for most purposes and the regulator must be capable of delivering steady pressures up to this maximum.

Precise control of carrier gas flow rate is important for several reasons: Retention time is flow-dependent; variation in flow rate will lead to a change in retention time and to uncertainty as to the identity of chromatographic peaks. Detector response is a function of flow rate so that quantitative precision is dependent upon the constancy of flow. Peak height is flow-dependent and thus the precision of peak height measurement is affected by variation in flow rate.

When operating under isothermal conditions, the use of a good flow control system and freedom from leaks in the system will ensure adequate constancy of flow rate. When, however, constant flow is required at different column temperatures or for temperature-programmed operation, some form of differential flow controller is required to compensate for changes in column back pressure with temperature. A typical controller is shown in Figure 4-2. If the pressure drop across the column changes, the diaphragm (A) will move, actuating the valve (B) so as to restore the pressure drop across the needle valve (C). Thus for a given setting of the needle valve, the flow rate will remain constant. Generally, the pressure

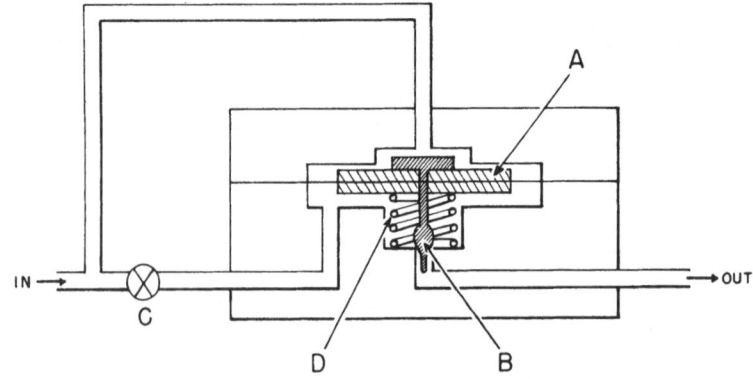

Figure 4-2. Differential flow controller.

upstream of the needle valve is set by the regulator on the gas supply cylinder, which must be capable of maintaining a constant pressure. The minimum pressure differential at which the controller will operate is determined by the spring (D). A differential of about 10 psi is usually required, so the supply pressure must be at least 10 psi greater than the maximum column back pressure. Typical commercial controllers will operate over a flow range of at least 0 – 200 ml/min, which is more than adequate for most needs.

A supplementary supply of carrier gas is sometimes needed to bring the flow of gas in the detector up to a value higher than that flowing in the column. Generally, this may be controlled by a needle valve, but if better control is needed a second differential flow control should be used.

Air and hydrogen supplies are required for operation of a flame ionization detector. Both may be controlled by the cylinder regulator in conjunction with a suitable restrictor in the gas supply line. A fine needle valve inserted in the line between the regulator and the restrictor will provide more precise control of hydrogen flow rate, should it be required.

INJECTION SYSTEMS

The sample to be introduced into the gas chromatograph is either a liquid (or solution) or a solid, and the primary requirement of the injection system is that the sample be vaporized as nearly instantaneously as possible so that a narrow band of vapor is introduced into the beginning of the column. If the concentration distribution of the sample vapor deviates too far from the ideal, that is to say, if the band is too broad or too asymmetric, column efficiency will suffer. It is also important that the injection system does not contribute to sample breakdown owing to reactive surfaces or hot spots in the injection port.

Whenever possible, it is preferable to inject the sample directly onto the column rather than into a chamber immediately preceding the column. On-column injection has the advantage that the sample comes into contact with the column only so that the possibility of breakdown through reaction with hot surfaces is minimized. In addition, the mass of hot material at the end of the column aids in transferring heat to the sample. Whatever system is used, it is important that the carrier gas be preheated to column temperature before it enters the column.

The size of sample used in steroid work is normally very small; volumes of a few microliters of solution or weights of a few micrograms of sample are typical. Thus, the problem of handling such small amounts of material must be solved. Three techniques are

described here; each of them has certan limitations but they represent
the best injection methods available at present.

Syringe Injection System

A typical syringe injection system is shown in Figure 4-3. The
sample contained in a small hypodermic syringe is injected into the
hot zone at the end of the column, where it is vaporized and swept
into the column by the carrier gas. The injection port is sealed by
a silicone rubber septum through which the hypodermic needle passes
when a sample is injected.

The system has the advantages of simplicity and universality.
Virtually every gas chromatograph manufactured is equipped with a
syringe injection port so that unless the user has special requirements,
he will use the syringe injection technique. The technique has several
general disadvantages. The range of sample volume that can be injected
is limited. Although syringes are available to deliver volumes between
0.1 and 1.0 μl, the precision of injection of less than about 0.3 μl is
poor. At the other end of the range, the extremely large solvent peak
produced by more than about 10 μl of solvent may obscure early sample peaks. It is difficult to handle volumes of solution of less than
about 50 μl; if only 10 μl or less of this solution can be injected into
the chromatograph, then at least a fivefold loss of sensitivity is
experienced. If very small amounts of material are to be estimated,
this loss of sensitivity could be serious. The syringes are relatively
expensive and fragile so that reasonable care must be taken in use.

In addition to the general disadvantages mentioned, there are certain problems connected with syringe injection systems that can be
solved by good design. The vaporization of the liquid sample produces

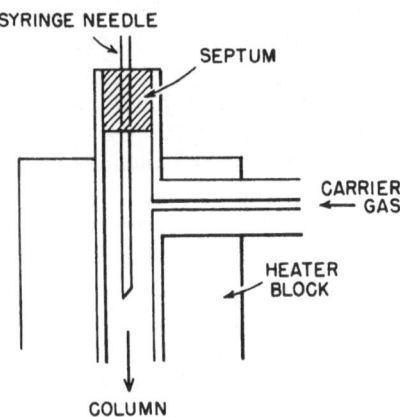

Figure 4-3. Syringe injection system.

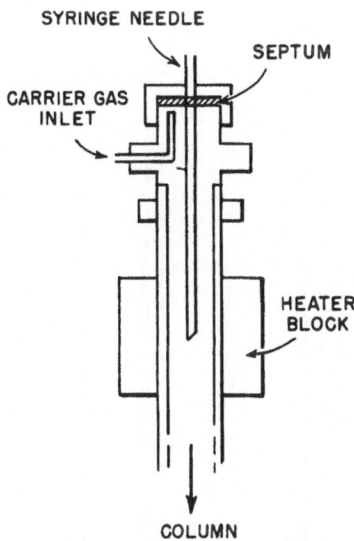

Figure 4-4. Improved syringe injection system. [Courtesy, F. & M. Scientific Corp.].

several milliliters of gas at the usual temperature of operation of the injection system and this rapid increase in volume produces an almost instantaneous pressure surge. Some of the vapor moves in a direction counter to the flow of carrier gas. Any volume in the system that is not swept by carrier gas, for example the volume between the septum and the carrier gas inlet tube, will fill with sample vapor. Any vapor trapped in this way will diffuse slowly back into the carrier gas stream and will increase the width of the sample band. Further-more, if any part of the system upstream of the point of injection is at a temperature much lower than that of the injection zone, sample will tend to condense and will then re-evaporate rather slowly. This effect will also contribute to band broadening.

A system described by Whittier and Umbreit [1] is shown in Figure 4-4. This system overcomes the problem of sample flashback by introducing the carrier gas to the injection zone through a small bore capillary. The high gas velocity and pressure drop through the capillary minimize the possibility of sample vapor traveling counter to the flow of carrier gas. In addition, the carrier gas sweeps the volume immediately adjacent to the injection septum, thus minimizing dead volume in this region.

The chromatograms in Figure 4-5 illustrate the difference between good and bad injection systems. The gain in efficiency resulting from the use of a good system is quite large.

It has been observed occasionally that material has been retained

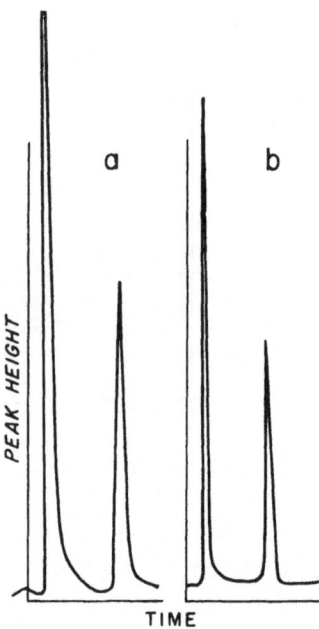

Figure 4-5. Effect of injection system on column performance.
(a) Simple system (Figure 4-3). (b) Improved system (Figure 4-4). [Courtesy, F. & M. Scientific Corp.].

on the septum and has later passed through the column, giving rise
to spurious peaks.   In the system described above, the septum is
swept by carrier gas and the possibility of adsorption of sample is
minimized.

Syringes
    The Hamilton microliter syringe has become the standard syringe
for gas chromatographic purposes and is described here. Other
syringes may be encountered occasionally, but since they operate on
similar principles most of what follows will be applicable.
    Two types of syringes are in general use. The first is a refined
and miniaturized version of a normal hypodermic syringe, illustrated
in Figure 4-6.   The second, illustrated in Figure 4-7, has a plunger
which extends into the syringe needle, the glass barrel serving only to

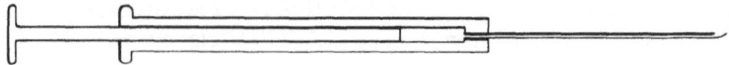

Figure 4-6. Ten microliter syringe.

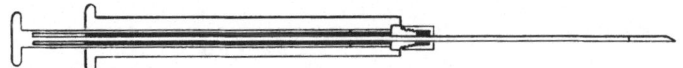

Figure 4-7. One microliter syringe.

measure volumes and to hold the parts of the syringe together. The second type is considerably more expensive than the first and we have found that it becomes difficult to achieve high precision of injection when very small volumes are injected (0.05 – 0.20 μl). Presumably some loss occurs from the needle tip; this forms an unacceptably large proportion of the total volume when the injection of very small volumes is attempted. In general, we prefer to arrange matters so that volumes of from 1 to 3 μl of sample are required and to use the conventional 10 μl syringe. With some experience and reasonable care, injection errors may be reduced to about $\pm 2\%$.

The needle of the syringe has a volume of about 0.6 μl so that even with the plunger fully depressed this amount of liquid remains in the syringe. The syringe needle becomes hot during sample injection and some of the liquid remaining in the needle will evaporate into the gas stream, the amount depending upon the length of time that the syringe remains in the injection port after the plunger has been depressed. The variation in the amount of residual sample evaporated constitutes the main source of error in the injection procedure and it can be minimized by adopting the following technique.

First, the syringe is filled with pure solvent and the plunger adjusted to the 1 μl mark. The needle is immersed in the sample and the requisite volume of sample drawn into the syringe. The volume of sample is measured directly by the difference in plunger positions when the syringe contains pure solvent only and when it contains solvent plus sample. It is important to ensure that no air bubbles are trapped in the syringe. The needle is inserted into the injection port, the plunger depressed fully, and the syringe withdrawn.

By this means, all the sample is injected into the chromatograph and only solvent remains in the syringe needle. Variation in the volume of solvent remaining in the needle will of course have no effect on the volume of sample injected. In addition to improving the precision of injection, the method is of value when it is imperative that all the sample be injected into the instrument as, for example, when radioactively labeled materials are being analyzed.

Tenney-Harris Injector System

This method, developed quite early [2], overcomes some of the objections to syringe injection. The injection of 0.1 – 0.2 μl samples

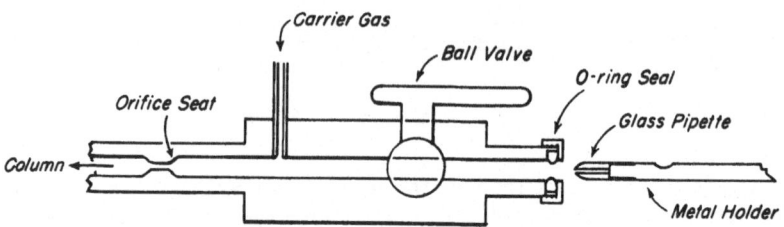

Figure 4-8. Tenney–Harris injection system.

with good precision is possible and, in general, plug injection is achieved. Disadvantages of the method are that a special injection system is required and injection directly onto the column is not possible.

The system is shown in Figure 4-8. The pipette is filled by capillary action when the tip is immersed in the sample solution. The outside of the tip may be wiped to remove excess sample, but care must be taken that liquid is not withdrawn from the capillary. The pipette is inserted into the injection port and the O-ring seal tightened. The ball valve is opened and the pipette pushed through the valve until it comes in contact with the orifice seat. At this instant a pressure drop is established across the capillary and the sample is swept into the column. The pipette is allowed to remain in contact with the orifice for a few seconds and is then withdrawn. Care must be taken to ensure that the tip of the pipette is clear of the valve before the valve is closed. The O-ring seal may then be loosened and the pipette withdrawn completely from the system.

Solid Injection System

A major disadvantage of liquid injection systems is that only a small proportion of the total sample is utilized. Several methods have been proposed in which comparatively large volumes of sample solution are taken, the solvent evaporated, and the solid residue injected into the chromatograph. A modification of the system of Menini and Norymberski [3] has given good results in our hands and is the best of the solid injection methods studied. Solutions of up to 1 ml in volume are readily handled, recoveries compare favorably with liquid injection, and most commercial instruments may be simply modified to accept the system.

The sample evaporation apparatus consists of a PTFE plate in which hemispherical depressions of approximately 2 cm diameter are formed. It is important that the surface of the depression be as smooth as possible, and Menini and Norymberski cold-worked the plastic using the rounded end of a heated Pyrex test tube; we have successfully employed methods better adapted to normal machine shop practice. The depressions are first machined using a $\frac{3}{4}$-in. ball end mill and are then lapped. The lapping tool consists of a $\frac{3}{4}$-in. steel ball joined to a length of heavy wall copper tubing by epoxy cement. Diamond lapping compound (5 μ grain size) is used and the lapping procedure repeated until the depression is seen to be smooth by visual inspection.

Sample collectors consist of small gauze cylinders approximately 3 mm long by 3 mm diameter formed from a 1 cm strip of 200-mesh stainless steel screen. They are washed thoroughly with chloroform before use.

The injection system is shown in Figure 4-9. The gauze cylinders holding the samples are contained in the side arm, which is sealed after loading with a silicone rubber injection septum. The side arm is attached to the vertical tube by a stainless steel Swagelok union to which a carrier gas inlet tube is attached. The vertical tube is attached to the column by a second Swagelok union.

A gauze cylinder is placed in each of the appropriate number of depressions in the plate, and the sample in c h l o r o f o r m solution is pipetted into each depression; up to 1 ml of solution may be used. The solution is allowed to evaporate; the chloroform solution does not wet the surface of the plate, with the result that the entire sample is

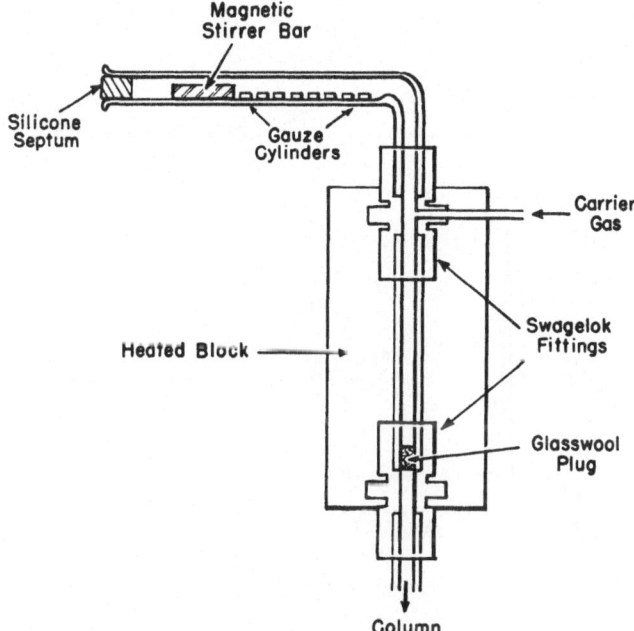

Figure 4-9. Menini–Norymberski injection system.

concentrated on the gauze cylinder. Here, it should be re-emphasized that it is of utmost importance that the plate surface be smooth. Any roughness in the surface will allow small drops of solution to become separated from the bulk with consequent loss of sample for the analysis.

When dry, the gauze cylinders are loaded into the side arm of the injection system, the stopper replaced, and the carrier gas flow allowed to attain equilibrium. The cylinders are pushed serially into the injection tube by means of the PTFE coated magnetic stirring rod, which is moved along the side arm by a small permanent magnet.

The chromatograms in Figure 4-10 of pregnanediol in urine compare syringe and solid injection methods. One μl of chloroform

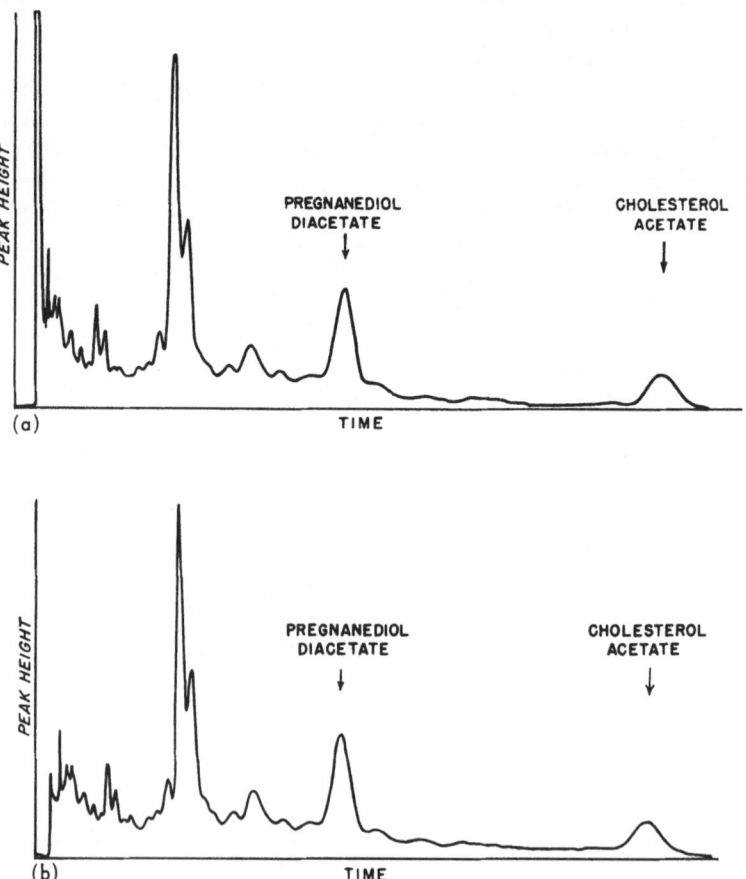

Figure 4-10. Comparison of (a) liquid and (b) solid injection systems. (See page 148 for analytical method.)

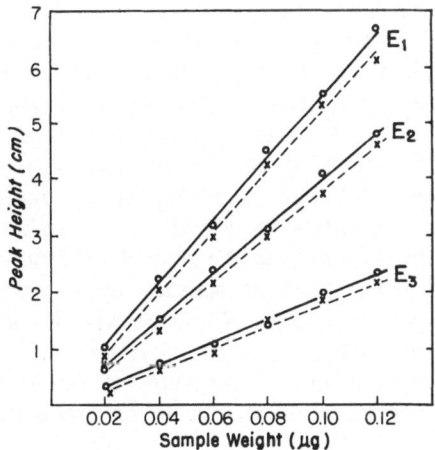

Figure 4-11. Comparison of recoveries. Solid line: Liquid injection. Dashed line: Solid injection. E_1: Estrone acetate. E_2: Estradiol diacetate. E_3: Estriol triacetate.

solution was taken for liquid injection, the same amount of material dissolved in 100 μl of chloroform for the solid injection. Calibration curves for estrogen acetates are shown in Figure 4-11. Losses from two main sources cause somewhat lower recoveries from the solid system: Approximately 1% of the material remains behind on the evaporation plate and the remainder appears to be lost on the gauze cylinders. No attempt has yet been made to reduce these losses since the results are satisfactory for most purposes.

COLUMN OVENS AND TEMPERATURE CONTROLS

The sole function of the column oven is to maintain constancy and uniformity of column temperature at the desired value. Constancy of temperature is important for reproducibility of retention times and for maintaining constant the rate of bleed of the column. Non-uniformity of temperature along the column normally leads to loss of column efficiency. Although unimportant from the point of view of analytical efficiency, it is convenient and time-saving if the column can be heated and cooled rapidly and if the column is readily accessible. For temperature-programmed operation, of course, the ability to change temperature rapidly is essential.

An air bath oven of low thermal mass with efficient air circulation seems to meet the above-mentioned requirements best and the majority of manufacturers have now adopted this design. This type of oven responds rapidly to changes in power input and a fully proportional temperature controller is necessary for proper operation. In such controllers, the actual oven temperature is continuously compared with that set on the controller. Any difference between the two is corrected by supplying the appropriate amount of power to the oven heaters. If the oven temperature is above the set point, no power is supplied until the oven has cooled to the set point. If the temperature is below the set point, power is supplied to heat the oven and as the oven temperature increases, the power input decreases. When the difference between oven temperature and set point is zero, the power input stabilizes at a value just sufficient to overcome thermal losses and to maintain the oven at the required temperature.

"On–off" controllers which operate by switching the power to the oven are not recommended. With this type of controller, the oven temperature cycles about a mean value and the column bleed rate follows this cycle. Change in bleed rate will often appear as a cyclic variation in the recorder trace, obscuring small peaks in the chromatogram.

Temperature control of other heated components in the system is usually less critical. Injection ports should be held at 10–20°C above column temperature; this can normally be done by surrounding the port by a metal block heated by a small cartridge heater. Control can be by means of an inexpensive thermostatic controller. The flame ionization detector, which is insensitive to small temperature changes, may be maintained slightly above column temperature in the same way. The electron capture detector, on the other hand, is relatively sensitive to temperature changes; its temperature can be held sufficiently constant by enclosing the detector in a heated metal block the temperature of which is controlled by a bimetallic thermostat.

DETECTORS

The advent of high-sensitivity ionization detectors made possible the development of practical methods for the gas chromatographic analysis of steroids. The early investigators mainly used the argon ionization detector [4–6], but experience has shown this detector to be insufficiently reliable for routine use. Its sensitivity is high, but it is too easily contaminated by column bleed and by analytical samples so that after a comparatively short time its behavior becomes erratic. The flame ionization detector [7–11], on the other hand, is inherently somewhat less sensitive but is far less prone to contamination problems. With reasonable care it is possible to operate a flame detector for many months and to achieve consistent results. Thus, for most purposes, the flame detector is the detector of choice. For certain applications, where extremely high sensitivity is required, the electron capture detector [12–14] may be employed. It cannot be emphasized too strongly, however, that this detector is very sensitive to changes in operating conditions and that great care is required for its successful operation. Thus, its use should be considered only when the flame detector proves inadequate.

Flame Ionization Detector

This detector was described almost simultaneously by McWilliam and Dewar [8] and Harley, Nel, and Pretorius [7]. The physical form of a typical detector is shown in Figure 4-12. Carrier gas emerging from the column is mixed with hydrogen and the mixture is burned at a small jet. Air is supplied to the detector, both to support the combustion of the flame and to sweep combustion products from the chamber. The electrode system impresses a DC field on the flame so that ions formed by the combustion process may be collected. The detector assembly is surrounded by a chimney which shields the flame from drafts. The detector is heated to a temperature sufficiently high to prevent condensation of sample in the connecting lines. A hot-wire ignition coil is mounted above the jet to provide a convenient means of flame ignition. A stable voltage source supplies the polarizing potential to the electrodes and the current flowing in the detector is displayed on a potentiometric strip-chart recorder which is connected to the detector by an impedance matching device, normally an electrometer amplifier.

Mode of Action of the Detector. When only carrier gas is flowing into the detector, flow of current is very small: that is to say, little ionization is occurring in the flame. Introduction of sample vapor mixed with carrier gas causes an increase in ionization and of current flowing in the external circuits. The nature of the processes leading to ion formation is obscure, but it seems likely that energy for

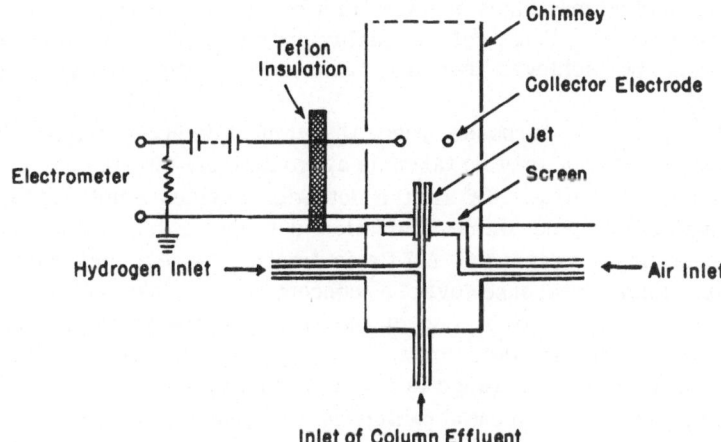

Figure 4-12. Flame ionization detector.

ionization is provided by the chemical reactions of the combustion process. A detailed discussion of flame chemistry is inappropriate here and the interested reader is referred to the many publications on the subject for more details [15–17].

Sensitivity of Detection. The sensitivity of any detection device may be defined in terms of the signal produced by a given amount of sample and of the noise level of the system. The ionization efficiency of the flame ionization detector is low; at best, only about one ion is produced per 10^5 molecules. However, the noise level of the detector itself is low so that it is possible to measure very small currents. Noise associated with small fluctuations in hydrogen and carrier gas flow rates forms a large proportion of detector system noise and for certain low-temperature applications extreme measures have been taken to control flow rates. Noise from these sources is insignificant compared with that associated with bleed-rate fluctuations of even the most stable high-temperature column. Thus, limit of detection in high-temperature operation is a function of the column and not of the detector.

Two other properties of the chromatographic column must be taken into account in order to arrive at a practical limit of detection for a particular analytical method. They are considered here since it is unrealistic to treat the detector in isolation from the rest of the system. First, sensitivity of detection is a function of retention time. The longer the sample zone takes to pass through the column the lower will be the concentration of the sample vapor in the carrier gas. Thus, for a given amount of material, the sensitivity will be greater at shorter retention times. In the same way, higher sensitivities will be achieved with more efficient columns. Second, losses of sample on the column may place a lower limit on the amount of material that can be detected. Thus, column quality not only defines the separation that can be achieved but may also control the sensitivity of the analysis.

For practical purposes, then, the limit of detection for the flame ionization detector may be taken as approximately 10^{-9} g.

Detector Linearity. Of all the detection devices employed in gas chromatography the flame ionization detector exhibits the widest range of linear response. All flame detectors within our experience exhibit linear response over a concentration range of 10^6:1; any nonlinearity at the lower end of the range can always be ascribed to causes other than the detector and, in the present context, nonlinearity at the upper end of the range is only of academic interest.

Effect of Applied Potential on Detector Response. Figure 4-13 illustrates the relationship between applied potential and current flowing in the detector. The plateau of the curve represents the region where

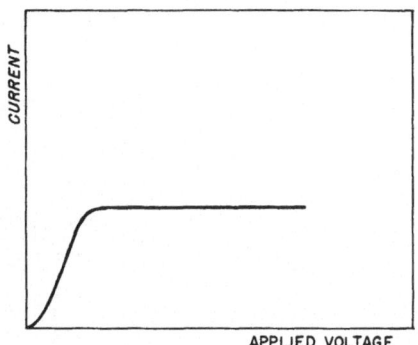

Figure 4-13. Relationship between applied potential and current
flowing in the detector.

all ions formed are collected. The extent of this region depends
upon the physical form of the detector, but it is always possible to
operate the detector far enough along the plateau so that small changes
in voltage have no effect on detector response.

Effect of Electrode Geometry on Detector Response. It is obviously important that all the ions formed in the flame are collected. Collection efficiency is a function of electrode geometry and applied potential and, usually, the user has no control over these factors. Examination of commercial flame detectors will demonstrate that a wide
variety of electrode geometries is employed, but in all cases good
collection efficiency is achieved. In some designs, electrode spacing
is rather critical and care should be taken to prevent mechanical
abuse of the electrode system.

Effect of Experimental Variables on Detector Response. If it is assumed
that high ion collection efficiency is achieved, the response of the detector is a function only of the number of ions produced and this in
turn is primarily a function of flame temperature. Variables under
the control of the operator that affect flame temperature are the nature
of the carrier gas, the flow rates of carrier gas, air, and hydrogen, and
the nature of the sample.

Carrier Gas. The effect of different carrier gases upon detector
response has been investigated by Sternberg [17] and Gudzinowicz [18].
The latter found that the relative response to a given compound was
1.00, 0.87, 0.50, and 0.12 for the gases argon, nitrogen, helium, and
hydrogen, respectively. Thus the response decreases with increasing
thermal conductivity and it is plausible to suppose that increase in
thermal conductivity results in a cooler flame and in decreased ion
production.

Gas Flow Rates. Above a certain minimum value, the flow rate
of air has little effect upon response. Once set, it is not necessary to

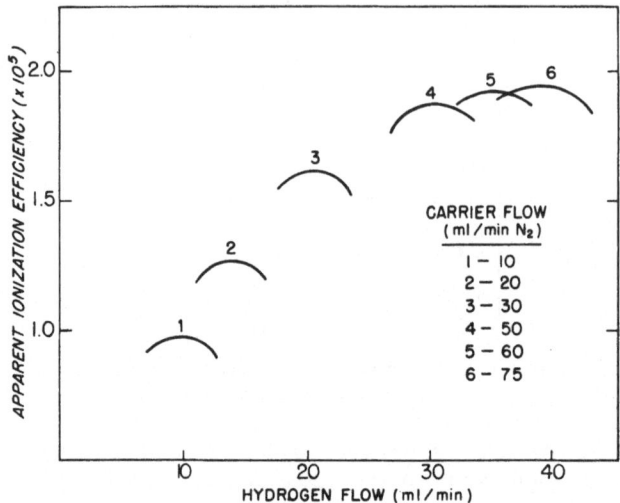

Figure 4-14. Effect of gas flow rates on detector sensitivity.

adjust the air flow.   The actual value required depends somewhat on the particular detector in use, but will be in the range 150–500 ml/min.

The effects of carrier gas and hydrogen flow rates are interrelated and must be considered together.   For any given carrier flow rate, the response exhibits a maximum as the hydrogen flow is changed; Figure 4-14 illustrates this relationship for a series of carrier flow rates.   It will be seen that detector response increases rapidly as carrier flow rate increases until a plateau is reached; thereafter, small changes in flow rate have little effect.   At very high flow rates, it is to be expected that response will again decrease since above a certain gas velocity ion collection efficiencies will decrease, but such flow rates are beyond the practical limits for analytical use.   The flow-rate range for maximum column efficiency is approximately 10–30 ml/min when small bore columns are used. At these flow rates, the detector is sensitive to changes in flow rate and sensitivity is considerably lower than optimum.   Both these problems can be overcome by adding additional carrier gas at a point between column and detector so that the detector is operating on the plateau shown in Figure 4-14. Modification of the fitting which joins column to detector normally provides a simple means of adding extra carrier gas.

Variation of Response with Molecular Species. The detector responds primarily to the carbon in the sample and for a homologous series at least, response is proportional to the number of carbon atoms in the molecule.   In general, hydrocarbons give the highest response; the presence of hetero-atoms tends to decrease the response.   Sufficient variation in sensitivity toward different molecules occurs so that for quantitative work it is necessary to calibrate for each individual compound.

The detector is insensitive to inorganic gases including water. Although it is of little direct interest to the steroid chemist, it should be noted that the presence of water may affect detector response, causing errors in quantitative work.

Detector Faults. Faults peculiar to a specific detector design are, of course, the province of the manufacturer. However, all detectors exhibit contamination effects, the seriousness of the problem being a function of the instrumental operating conditions rather than of the design of the detector. In general, the sample contributes little directly to detector contamination. The major source of contamination is the column, either from bleeding of stationary phase or, to a lesser extent, from the slow elution of sample decomposition products that have accumulated on the column.

Contamination is likely to occur at three places in the detector: the line conducting the sample from the column to the detector jet, the detector jet itself, and the electrode system. Condensation of stationary phase bleed in the sample line results in the formation, in effect, of a short inefficient chromatographic column. Thus, the effect manifests itself as a loss in column efficiency and sometimes as a loss in sensitivity. The problem can be minimized or prevented entirely by maintaining the system at a temperature high enough so that condensation does not occur. Once the line is contaminated it may be cleaned by raising its temperature or, if this is not successful, by washing out with a suitable organic solvent.

Eluant from the column may lodge in the detector jet, partially blocking it. When this occurs, the flame will usually behave erratically; it may prove difficult to keep the flame lit without increasing hydrogen flow rate excessively and the system noise level will increase. Inspection of the jet will usually reveal contamination since blocking occurs most frequently at the tip. The jet should be changed or, if fixed, it may be cleaned by passing a stiff wire through the bore.

Electrode contamination occurs primarily as a deposition of silica, a combustion product of silicone stationary phases. Unless the buildup of silica is very great, it is unlikely to affect electrical performance of the detector. However, particles of silica becoming detached from the electrodes may fall into the flame, giving rise to noisy operation. In some designs of detector, it may be possible for silica to coat insulators holding signal electrodes. Water absorbed by the silica may lead to partial breakdown of the insulator and an increase in noise level and standing current.

Electron Capture Detector

Although of much less general application than the flame ionization detector, the electron capture detector has proved to be of value in those situations where extremely high sensitivity is required. The detector must be operated with great care if optimum results are to be obtained and for this reason a detailed discussion of its operating characteristics is included here.

The physical form of the detector is simple (Figure 4-15). The electrode system consists of two parallel disks of stainless steel or brass separated by a PTFE insulator. The sample enters through a tube leading to the anode; this electrode is covered by a fine mesh screen which serves to produce uniform gas flow in the detector chamber. The gas flow passes through the chamber and is exhausted through a hole in the cathode. The primary source of ionizing radiation is attached to the cathode and consists of a disk of stainless steel coated with titanium tritide. The activity of the tritium is usually of the order of 100–300 mC. The external circuitry consists of a polarizing source for the detector and an electrometer amplifier. Like the flame detector, the electron capture detector is a high-impedance device so that a good quality impedance matching device is required to convert the detector output to a form acceptable to a strip-chart recorder. The detector may be polarized by applying either a steady negative voltage or a square microsecond pulse to the cathode.

Electron Capturing Processes. When only carrier gas is flowing through the detector, the electrons emitted by the radioactive source react with the carrier gas molecules to produce positive ions:

$$N_2 + e^- \text{ (energy } E) \to N_2^+ + e^- + e^- \text{ (energy } < E)$$

This process continues until the primary electron has insufficient energy to produce further ion pairs. Under the steady conditions obtaining during the operation of the detector, the chamber will have a steady concentration of both positive ions and electrons. If a potential

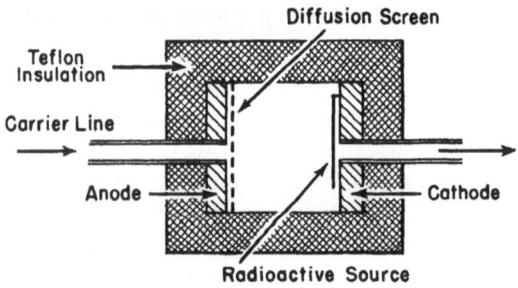

Figure 4-15. Electron capture detector.

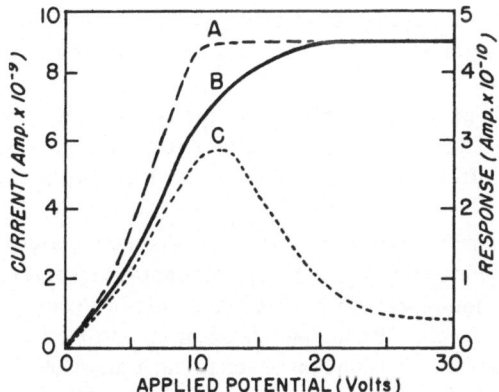

Figure 4-16. Electron capture detector. A, B: Current–voltage relationships. C: Sensitivi-ty–voltage relationship. [S. J. Clark, Residue Reviews, 5, 32 (1964)].

is applied across the cell, the charge carriers will be collected at the electrodes and a current will flow in the external circuit. A plot of applied potential vs. current has the form shown in Figure 4-16. The dotted curve A is the theoretical one calculated from considerations of detector geometry and the randomness of the ionization process. The solid curve B is obtained experimentally and the deviation from theory is a measure of the extent of the space charge set up in the detector.

If now a sample, capable of capturing electrons, is introduced into the detector, mixed with the carrier gas, electron attachment will occur:

$$XY + e^- \begin{array}{c} \rightarrow XY^- \\ \searrow X + Y^- \end{array}$$

Electron attachment is a complex process, but two main mechanisms seem to predominate: resonance capture, akin to light absorption, in which the reaction product is an excited but undissociated negative ion, and dissociative capture, where dissociation of the ion occurs to give a free radical and a negative ion.

The net effect of the capturing process is to substitute a relatively slow-moving negative molecular ion for a highly mobile electron. The reduction in negative ionic mobility is detected either through ion recombination in the DC mode of operation or directly in the pulsed mode. The lower mobility of the negative molecular ion increases greatly the possibility of ion recombination:

$$N_2^+ + Y^- \rightarrow N_2 + Y$$

with the result that charges are neutralized and the current flowing in the detector decreases. In the pulsed mode of operation, the pulse is sufficiently short so that only electrons are collected and, again, the formation of negative ions results in a decrease in the current flowing in the detector.

For a given compound, the response of the detector depends upon the extent to which negative ions are formed or lost; any change in conditions affecting such processes will lead to a change in sensitivity.

Effect of Experimental Variables Upon Detector Performance

Applied Potential. At very low applied potentials, separation of the positive ion—electron pair is less likely to occur and recombination can take place. Response to sample entering the detector is a function of free electron concentration in the ionization chamber; hence, sensitivity would be expected to increase in proportion to the standing current as potential is increased, reaching a maximum at saturation. The third curve in Figure 4-16 illustrates the relation between voltage and response; it is seen that, in fact, response reaches a maximum at approximately that voltage theoretically required for saturation. The presence of sample molecules apparently reduces the effect of the positive ion space charge, through molecular ion recombination, allowing maximum response at a voltage lower than would have otherwise been expected.

At potentials greater than that required for maximum response, the charge carriers in the chamber are further accelerated, thus reducing the probability both of electron absorption and of ion recombination. In this region, sensitivity is approximately inversely proportional to the square of the applied potential.

Pulsed Operation. In this mode of operation, highest sensivity is attained when the maximum time is provided for reaction to occur: that is, using a pulse of minimum width at a low repetition rate. At pulse widths and repetition rates below a certain value, electrons are lost through recombination and at the walls of the ionization chamber. Thus, response will exhibit a maximum for certain experimental conditions. A pulse repetition rate of 10 KHz and a width of $1\,\mu$sec have been found to give maximum sensitivity. Response is decreased at higher or lower pulse repetition rates. At a repetition of 10 KHz, pulse width has no effect on sensitivity in the range 1–3 μsec. The detector is insensitive to pulse amplitude in the range 35–90 V; below 35 V sensitivity is decreased. Pulse shape does not appear to be critical; however, a generator providing reasonably square pulses is recommended since the limit beyond which pulse shape affects response has not yet been determined.

Carrier Gas Flow Rate. Figure 4-17 shows current-voltage relationships for a number of flow rates. A change in gas flow rate will affect the drift velocity of ions in the chamber, and the geometry of

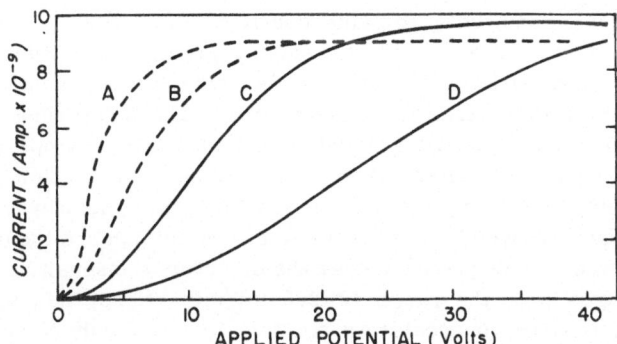

Figure 4-17. Electron capture detector.   Effect of temperature and flow rate on detector standing current. N_2 carrier gas. [S. J. Clark, Residue Reviews, 5, 32 (1964)].

	Detector temperature, °C	Flow Rate, ml/min.
A	220	160
B	220	40
C	120	160
D	120	40

the chamber is such that positive ions move in the same direction as the carrier gas stream.   Thus, an increase in flow rate will reduce the randomness of ionic velocity and saturation will occur at a lower applied potential.   For each flow rate, a response curve can be constructed exhibiting maximum sensitivity at approximately theoretical saturation potential for that flow rate.   The relationship between flow rate and potential for maximum response is shown in Figure 4-18. Sensitivity at maximum response is constant over the flow-rate range 15–200 ml/min and occurs always at the same value of standing current.   Thus, provided that voltage is adjusted to compensate for changes

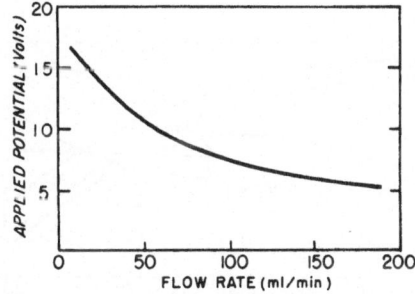

Figure 4-18. Electron capture detector. Relationship between potential for maximum response and flow rate. N_2 carrier gas.   [S. J. Clark, Residue Reviews, 5, 32 (1964)].

in flow rate, sensitivity remains constant. However, it is obviously essential for accurate work to employ precise flow control and if temperature-programmed operation is contemplated, some form of differential flow regulation is mandatory. The deleterious effect of poor flow control can be alleviated to some extent by operating the detector at high flow rates, where it is less sensitive to flow changes. If the column must be operated at low flow rates, additional carrier gas may be introduced at a point between the column and the detector.

Operated in the pulsed mode, the detector is insensitive to flow-rate changes over the range 40–200 ml/min. Here, the pulse dura-tion is too short to accelerate positive ions and only electrons are collected. Hence, changes in the motion of positive ions will not be observed.

Detector Temperature. Movement of the ions in the cham-ber is temperature-dependent so that changes in detector temperature in a DC polarized system will affect potential for maximum response in much the same way as do changes in flow rate. The effect upon stand-ing current is illustrated in Figure 4-17. For pulsed operation, little change in standing current is observed over the range 110–225°C. Presumably, electron mobility is so high that it is relatively un-affected by temperature changes.

All of the reactions occurring in the ionization chamber are temperature-dependent so that a temperature effect in addition to that on ionic mobility would be expected. The series of reactions occurring in the detector is far too complex to allow, for the present, quantita-tive evaluation of the temperature dependence of each. It seems likely, however, that apart from the electron attachment reaction, tempera-ture coefficients will be of approximately the same magnitude for all compounds.

Dependence of sensitivity upon temperature for two different types of strongly capturing compounds is illustrated in Figure 4-19. It will

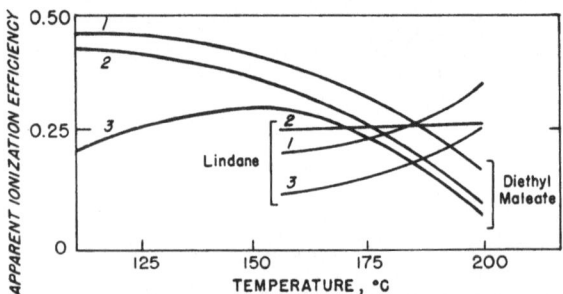

Figure 4-19. Electron capture detector. Dependence of sensitivity upon temperature. Carrier gas flow rate: 150 ml/min. (1) DC. argon–methane; (2) Pulse. argon–methane; (3) DC. nitro-gen. [S. J. Clark, Residue Reviews, 5, 32 (1964)].

be seen that change of response with temperature is strongly dependent
not only upon compound type but also upon the mode of operation of
the detector.   Temperature response curves for the less strongly
capturing compounds anthracene and 1,2,3,4-tetrachlorobenzene have
also been measured.   The curves for anthracene have the same
general form as those for diethyl maleate, whereas 1,2,3,4-tetra-
chlorobenzene exhibits yet a third type of behavior. Here, sensitivity
decreases by approximately 20% over the range 110–160°C, remaining
almost constant thereafter. For both compounds, sensitivity is affected
by operating conditions.   Highest sensitivity is achieved with argon-
methane carrier and DC operation, lowest with nitrogen carrier and
DC operation, and an intermediate value is obtained with pulse
operation.

It is evident that the temperature dependence of the detector is
extremely complex and much more experimental work would be required before any general statement could be made relating, for
example, electron temperature to sensitivity. However, several important practical consequences emerge: First, the detector must be
maintained at a fixed temperature independent of the remainder of the
instrument. Second, comparison of results from different sources will
be fruitless unless the temperature dependence of the system is established, or unless experiments are universally conducted at a fixed
temperature.   For steroid work it is suggested that the detector be
maintained at 210°C; this value is sufficiently high to minimize contamination from column bleed, yet low enough so that loss of tritium
from the radioactive source is not serious. Third, the effect may be
used to advantage either for qualitative work or to select conditions
such that interference with the compound sought is minimized.   Finally,
caution must be exercised in comparing results obtained with DC
operated detectors having differing geometries. Here, the distribution
of electron energies is affected not only by temperature changes, but
by the uniformity or otherwise of the potential gradient across the
ionization chamber.

Sensitivity of Electron Absorption Measurement.
The ultimate sensitivity of any ionization detector is a function of fluctuations in the detector standing current. For an electron capture detector having a saturation current of $1 \cdot 10^{-8}$ A, the noise associated
with the ionization process is approximately $5 \cdot 10^{-14}$ A and the corresponding limit of detection approximately $5 \cdot 10^{-19}$ mole/sec. In practice, noise from other sources greatly exceeds that from the ionization
process so that in a practical system, a noise level of the order of
$1 \cdot 10^{-12}$ A may be expected.   Thus, a practical limit of detection is
approximately $1 \cdot 10^{-17}$ mole/sec and this will be further reduced
since ionization of the sample is rarely 100% efficient. As for the

flame ionization detector, the practical limit of detection is usually set by experimental conditions. For the most strongly capturing compounds, a limit of detection of approximately 10^{-12} g can be achieved.

Linearity of Response. Over approximately 40% of the current range of the detector response at optimum sensitivity is related to sample weight as follows:

$$I = I_0 e^{-kc}$$

where I is current flowing in the detector at sample peak maximum, I_0 is detector standing current, c is sample weight, and k is a constant. As Lovelock [14] has pointed out, the relationship bears a formal analogy to Beer's law or to the concentration relationship for a second-order reaction.

Nature of the Carrier Gas. Argon containing 10% methane is the most suitable of the carrier gases so far studied for operation in the pulsed mode; nitrogen is commonly used for DC operation.

For the DC mode, sensitivity is greater by a factor of approximately 1.3 when using argon–methane. The difference can be accounted for by the fact that the saturation current for a given detector is higher by a similar amount when the mixed gas is employed. There is comparatively little difference in sensitivity between DC and pulsed systems when both employ argon–methane.

Electron mobility is higher in the argon–methane mixture than in nitrogen and, consequently, lower polarization voltages are required for the former gas. Flow and temperature effects connected with ionic mobility are qualitatively the same for both gases; however, the magnitudes of the effects are greater for nitrogen.

Comparison of DC and Pulse Systems. From a theoretical point of view [14], there can be little doubt that electron absorption measurements are best made using thermal electrons under zero field conditions. At the present stage of development of instrumentation, this is most easily done by applying short pulses to the detector, thereby to sample the electron population in the ionization chamber. From the practical standpoint, it is necessary to inquire into the relative merits of pulse and DC systems in order to determine whether or not the extra cost and complexity of the pulse system is worthwhile.

The sensitivity of the detector is approximately the same for both systems. What difference exists can be attributed to the nature of the carrier gas used; nitrogen is generally employed with DC polarized detectors whereas argon–methane is the preferred carrier gas for pulsed operation. If the highest possible sensitivity is required from a DC detector, there is no reason why argon–methane should not be used although it is more difficult to obtain and more expensive than nitrogen.

Insensitivity to changes in carrier gas flow rate represents a real advantage of pulsed operation. If, however, a DC detector is operated in conjunction with a suitable flow controller and at a reasonably high flow rate, errors from this source are minimal.

Adjustment for optimum linearity is more critical in the DC system and spurious results are obtained if the applied voltage is too high. The ability to change sensitivity by a factor of approximately ten is an advantage of the pulsed system; this can be done during an analysis, although readjustment of the baseline will then be necessary. In many circumstances, dilution of the sample with a transparent (noncapturing) solvent will serve the same purpose.

Insufficient work has yet been done on temperature effects to assess the relative merits of the two systems, but it appears that the pulsed system is less temperature-dependent. If adequate temperature control of the detector is available and it is operated at a fixed temperature, little trouble should be experienced.

Lovelock [14] has recently published an authoritative review of the operation of the electron absorption detector and described in detail possible sources of error associated with DC operation of the detector. In practice, it has been found possible to operate the detector over periods of many months without trouble from any of these effects. It must be emphasized, however, that great care must be exercised in the operation of the DC system and that from the standpoint of ease of operation, the pulse technique is to be preferred. Otherwise, DC operation is perfectly adequate for use in the determination of steroids.

Effect of Molecular Species on Response. Only those compounds capable of capturing electrons give rise to an appreciable response in the detector; for a very strongly capturing compound, the response is of the order of 10^6 times greater than for an aliphatic hydrocarbon, for example [19]. Heavily halogenated compounds capture strongly; this property can be exploited in steroid gas chromatography by the preparation of halogenated derivatives such as the heptafluorobutyrates or monochloroacetates. Worth noting, although of less immediate practical interest, is the behavior of certain free steroids. In general, steroids with the Δ^4-3-one structure capture electrons, the extent to which capture occurs depending also on other substituent groups in the molecule. Lovelock [20] has discussed the capturing properties of such steroids in relation to their biological activity.

Detector Faults. Since the detector is highly sensitive to changes in operating conditions, it is essential that temperature and flow controls are adequate and operating properly. This being so, those faults that can be corrected by the operator are concerned exclusively with the presence of impurities in the gas stream entering the detector.

First, it is essential that the carrier gas be free from moisture and oxygen since those impurities will cause partial or complete loss of detector sensitivity. Use of a good grade of dry carrier gas will normally ensure absence of water, but, as an added precaution, the use of a molecular sieve trap in the gas line is recommended. Presence of air is usually the result of small leaks in the system; in certain configurations, a Venturi effect occurs with the result that air is drawn into the gas line.

Of even greater consequence is the effect of column bleed upon detector sensitivity, illustrated in Figure 4-20. Every precaution must be taken to ensure that column bleed is minimized by long conditioning and, although it is sometimes necessary to use comparatively unstable columns, it must be recognized that this practice will inevitably cause impaired detector performance. Accumulation of sample breakdown products on the column invariably occurs to a greater or lesser extent. This material is eventually eluted from the column with the same effect as column bleed. A contaminated column can usually be improved by reconditioning.

In addition to loss of sensitivity, a problem common to all electron capture detectors, gradual contamination of the source from deposition of stationary phase occurs in those detectors employing tritium sources. Contamination may be recognized both by decrease in detector standing current and sensitivity, and by distortion of the chroma-

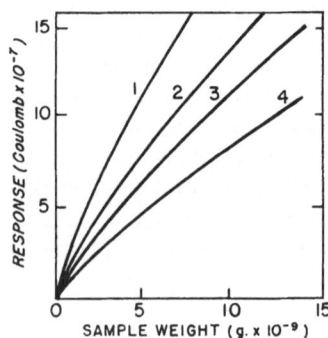

Figure 4-20. Electron capture detector. Effect of column bleed upon sensitivity. Column: 10% DEGS on 80-90 mesh Anakrom A. 110°C. Flow rate: 150 ml/min. [S. J. Clark, Residue Reviews, 5, 32 (1964)].

Unconditioned column	Conditioned column
(2) Argon—methane; pulse	(1) Argon—methane; pulse
(4) Nitrogen; DC	(3) Nitrogen; DC

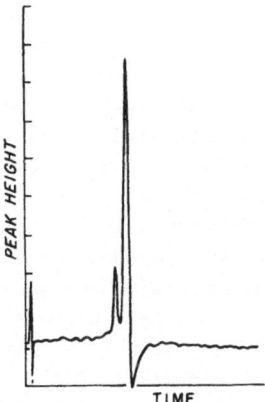

Figure 4-21. Electron capture detector. Peak distortion arising from detector contamination.

tographic peak (Figure 4-21). The reason for this latter effect is not entirely clear, but it is probable that, during passage through the detector, the sample is sorbed and desorbed by the layer of material on the source and that the electrical properties of the detector change as a result. If contamination is not too severe, operation of the detector for a period of 48 hr with a flow of hot pure carrier gas or flushing the c o l d detector with organic solvents may improve the condition. Otherwise, it is necessary to return the detector to the manufacturer for replacement of the source.

ELECTROMETER AMPLIFIERS

The primary purpose of the electrometer amplifier is to match a device of very high impedance, the detector, to the comparatively low impedance of the recorder input network. Subsidiary functions are to cancel out any steady currents generated in the detector that are not associated with signal due to the sample, and to provide attenuation of the signal so that a wide current range may be measured.

Functionally, the electrometer must be linear and stable since these properties affect the accuracy and precision of the analysis. Generally speaking, the maximum useful amplifier sensitivity is of the order of 10^{-11} A. That is, a current of 10^{-11} A flowing in the detector will cause a full-scale deflection of the recorder pen. Many instruments provide for higher sensitivities, but these are often unusable owing to the noise generated in other parts of the apparatus which, of course, is amplified to the same extent as the signal from the sample. For steroid work, at least, it is unlikely that sensitivities lower than 10^{-8} A will be required although many amplifiers provide ranges down to 10^{-5} or 10^{-6} A.

Voltage supplies for the detectors are associated with the electrometer amplifier for the reason that it is usually possible to derive stable DC voltages from the electrometer power supply. In some units batteries are used; these constitute perfectly satisfactory and stable voltage sources providing that there is no objection to the slight inconvenience of changing them at regular intervals. The flame ionization detector is always operated in such a manner that it is insensitive to small changes in voltage. Thus, some fluctuation in the voltage supply can be tolerated. The DC operated electron capture detector, on the other hand, is quite sensitive to voltage changes and under the most sensitive conditions a change of 0.1 V will produce a measurable change in detector sensitivity. In addition, the supply for this detector must be variable. Pulse generators for operating this detector are supplied by some manufacturers. They normally operate at a fixed amplitude and pulse width, but control of pulse repetition rate should be provided since this gives a very convenient means of changing detector sensitivity.

RECORDERS

Little need be said about recorders since most manufacturers supply a suitable recorder with each instrument. As long as the output from the chromatograph electronics matches the input sensitivity of the recorder, virtually any recorder from a reputable manufacturer can be used with any chromatograph. A useful discussion of the properties of recorders which affect analytical performance was published recently by Aaker [20a].

CONNECTION BETWEEN COMPONENTS OF THE CHROMATOGRAPH

Steroids are chromatographed well below their boiling points and a significant drop in temperature anywhere in the chromatographic system will lead to condensation of sample. It is essential, therefore, thay any connection between components, e.g., between column and detector, be maintained at the proper temperature. Furthermore, connectors having any appreciable dead volume will cause loss of separation efficiency. Thus, any long connecting lines are suspect and it is better if all components in the analytical system are joined directly one to another.

REFERENCES

1. M. B. Whittier and G. R. Umbreit, Facts and Methods for Scientific Research 6:3 (1965).
2. H. M. Tenney and R. J. Harris, Anal. Chem. 29:317 (1957).
3. E. Menini and J. E. Norymberski, Biochem. J. 95:1 (1965).

4. J. E. Lovelock, J. Chromatog. 1:35 (1958).
5. J. E. Lovelock, A. T. James, and E. A. Piper, Ann. N.Y. Acad. Sci. 72:720 (1959).
6. J. E. Lovelock, Gas Chromatography, 1960, edited by R. P. W. Scott (Butterworths, London, 1961), p. 16.
7. J. Harley, W. Nel, and V. Pretorius, Nature 181:177 (1958).
8. I. G. McWilliam and R. A. Dewar, Nature 182:1664 (1958).
9. I. G. McWilliam and R. A. Dewar, Gas Chromatography, 1958, edited by D. H. Desty (Butterworths, London, 1959), p. 142.
10. L. Onkiehong, Gas Chromatography, 1960, edited by R. P. W. Scott (Butterworths, London, 1961), p. 7.
11. R. D. Condon, P. R. Scholly, and W. Averill, Gas Chromatography, 1960, edited by R. P. W. Scott (Butterworths, London, 1961), p. 30.
12. J. E. Lovelock and S. R. Lipsky, J. Am. Chem. Soc. 82:431 (1960).
13. J. E. Lovelock, Anal. Chem. 33:162 (1961).
14. J. E. Lovelock, Anal. Chem. 35:474 (1963).
15. H. F. Calcote and I. R. King, Symp. Combust., 5th, Pittsburgh, 1954 (1955) (Reinhold, New York, 1955), p. 423.
16. H. F. Calcote, Combust. Flame 1:385 (1957).
17. J. C. Sternberg, W. S. Gallaway, and D. T. L. Jones, Gas Chromatography, edited by Brenner, Callen, and Weiss (Academic Press, New York, 1962), p. 231.
18. B. J. Gudzinowicz, private communication.
19. J. E. Lovelock, Nature 189:729 (1961).
20. J. E. Lovelock, P. G. Simmonds, W. J. A. VandenHeuvel, Nature 197:249 (1963).
20a. A. A. Aaker, Anal. Chem. 37:1252 (1965).

Chapter 5

Ancillary Methods

Gas chromatography is among the most powerful of separatory methods available to the chemist and, for most purposes, it is a sufficiently precise means of quantitative analysis. However, it is a relatively poor qualitative tool. Even when techniques such as those described in Chapter 13 are used, identification of unknown compounds is relatively uncertain. Although identification methods are of little interest to the laboratory exclusively engaged in routine analysis, it is of utmost importance during the development of methods not only that the compounds of interest be thoroughly characterized but that the identity of the individual chromatographic peaks be unequivocally established. The elution of more than one component at the same time is relatively common and, obviously, the presence of any impurity in a peak of interest will affect quantitative accuracy. Furthermore, it is important to determine whether decomposition or rearrangement of the compounds of interest has occurred in the chromatographic system. Information of this sort can only be obtained by subjecting the components separated in the chromatograph to further analysis; in this context the gas chromatograph can be regarded as a purification method for the particular analytical technique to be employed.

TRAPPING METHODS

The eluate from the chromatographic column may be dealt with in two general ways: It may be led directly to the apparatus to be used for further analysis, or fractions of interest may be collected to await subsequent analysis. Fractions are collected by cooling the gas stream emerging from the column and condensing the fractions in suitable vessels. Provided that certain precautions are taken, the technique is very simple and high recoveries of individual compounds may be achieved. The main problem in the condensation of high-molecular-weight materials such as steroids, from low-molecular-weight carrier gases, lies in the tendency of such mixtures to form fogs when cooled. When a fog forms the droplets are swept through the receiving vessel and most of the sample is lost.

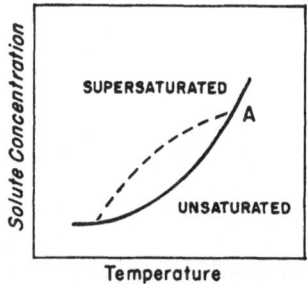

Figure 5-1. Vapor pressure relationship for solute–carrier gas mixture.

The state of the gas–vapor mixture may be illustrated by a vapor pressure diagram (Figure 5-1). At some point during the cooling process, the carrier gas–sample mixture will become saturated with respect to the sample (point A). If further cooling results in supersaturation of the carrier gas, a temperature will be reached at which rapid condensation and fog formation will occur.

Condensation depends upon the simultaneous flow of heat and of mass (sample) toward the wall of the tube, and thus upon the thermal diffusivity of the mixture and the (mass) diffusion constant of the sample molecules in the carrier gas. For large molecules, the flow of heat is always faster than the diffusion of the sample molecules and fog is bound to form (dotted curve in Figure 5-1).

Several methods are available to prevent fog formation; the simplest of these is to ensure that cooling is gradual so that at each point on the vapor pressure curve the system has time to come to equilibrium. Fortunately, the vapor pressure of steroids is sufficiently low at or near room temperature so that extra cooling is unnecessary and all that is required is that a suitable temperature gradient be set up along the sample collection tube. A practical system is shown in Figure 5-2 [1]. A fresh length of capillary tubing is inserted for each fraction of interest. A capillary tube must be kept in the outlet at all times to prevent contamination of the system by effluent even if samples are not to be collected.

Since the flame ionization detector consumes the sample, arrangements must be made to split the effluent stream from the column so that the minimum amount of material required for detection is fed to the detector and the remainder to the sample collector.

QUALITATIVE ANALYSIS

Qualitative methods should be reasonably rapid, selective, and applicable to a wide range of compounds. Because of the power and versatility of spectroscopy, one or another of the spectroscopic

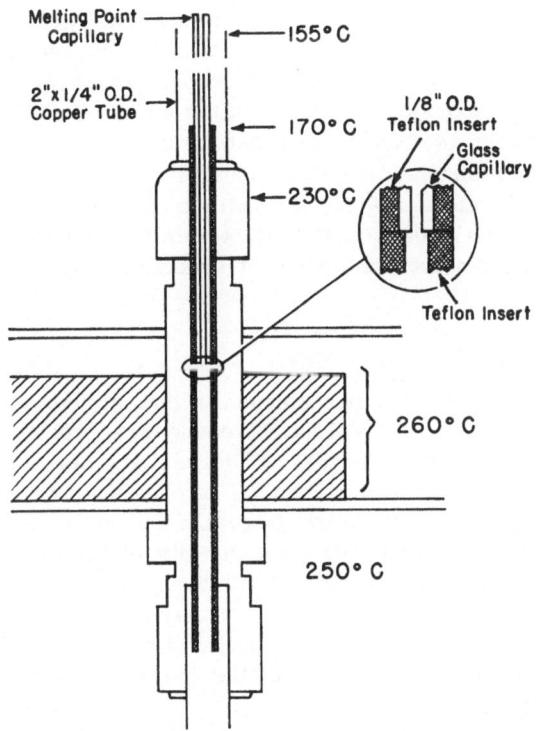

Figure 5-2. Trapping system. [Courtesy, F. & M. Scientific Corp.].

techniques is likely to be the main tool for qualitative analysis. What follows is a necessarily brief comparison of the available methods, giving some indication of the merits and shortcomings of each.

Ultraviolet Absorption Spectrometry

This has been employed longer than any other spectroscopic method, having first been used about 1930. The subject has been thoroughly reviewed by Dorfman [2] and Bernstein [3]. Generally, steroids containing nonconjugated double bonds or keto groups exhibit either weak absorption or absorption at very short wavelengths. In some cases, derivative formation can be used to enhance absorption or to induce wavelength shifts. Molecules containing conjugated systems exhibit strong absorption, and the correlation of both band intensity and wavelength with structure has resulted in the widespread use of ultraviolet absorption as a means of characterization. Uncertainty of identification can arise from the fact that many steroids exhibit similar spectra, differing only in the intensity of the absorption bands. For the same reason, detection of small amounts of impurities in chromatographic fractions or the recognition of mixed peaks may be difficult or impossible. Samples of the order of 10^{-4} to 10^{-6} g are required to obtain good spectra so that the method is only moderately sensitive by gas chromatographic standards. Equipment for measuring spectra down to about 200 mμ is readily available and is moderately expensive ($5000–15,000). As yet, techniques have not been developed to allow direct recording of spectra of fractions emerging from the column.

Optical Rotatory Dispersion

The application of the phenomenon of optical rotatory dispersion to the elucidation of structural and stereochemical problems in organic chemistry was pioneered largely by Djerassi and his co-workers [4]. The special properties of the steroidal ketones made the study of this group of compounds particularly favorable and much information has been gathered. The conformational rigidity of the structure allows unambiguous determination of configuration. The ideal nature of the keto group as a chromophore, the presence of a large number of asymmetric carbons, and the availability of many positional isomers made possible the orderly study of the interaction of chromophores with centers of asymmetry.

The technique is valuable since it allows the determination of absolute configuration. However, rather large samples are required (10^{-4} to 10^{-3} g) and suitable equipment is quite expensive. It is unlikely that optical rotatory dispersion will provide the main tool for characterization.

Infrared Spectrometry

Infrared spectrometry is undoubtedly the most widely used of all physical methods for the characterization of organic compounds. As in optical rotatory dispersion, the steroids have been studied in detail and a large catalog of spectra is available [5]. Thus the task of the experimenter is greatly simplified, since infrared spectrometry, like so many physical methods, depends for its success upon the availability of reference spectra.

Instrumentation is widely available and it is the rare laboratory that does not either possess or have access to an infrared spectrometer. The trapping of gas chromatographic fractions for infrared study is now common practice and effective techniques have been developed [6]. Trapping accessories are available from the manufacturers of infrared instrumentation. Recently, instrumentation for the direct recording of spectra of the effluent from the chromatograph has become available [7], but its effectiveness in handling high-boiling materials has yet to be assessed.

Although, in special circumstances, spectra can be obtained from a microgram or so of material, 10^{-5} to 10^{-4} g is usually required. Infrared spectra are not normally sensitive to small amounts of impurities so that the presence of a few percent of an impurity in a fraction may not be detected. On the other hand, major impurities are readily detected.

Nuclear Magnetic Resonance Spectrometry

The technique of nuclear magnetic resonance spectrometry is in a state of rapid growth; during the past two or three years many papers have been published on its application to the structural determination of organic compounds. Much information of interest to the steroid chemist is to be found in the literature. It is unfortunate that often, when the technique is not of primary interest, no mention of its use is made either in the title or abstract of a paper. However, the recent introduction of an abstract service [8] should go far in remedying the situation. The reviews of Jardetzky and Jardetzky [9], Stothers [10, 11], and Bhacca and Williams [12] form a useful introduction to the subject. A technique for the direct trapping of a gas chromatographic fraction in a NMR cell has been described recently [13]. Such a method used in conjunction with one of the newer spectrometers and computer averaging techniques allows a good spectrum to be obtained from samples of the order of 10^{-6} g. During the past few years instrumentation has improved enormously so that, despite its cost and complexity, the routine determination of spectra is now a relatively simple matter.

Mass Spectrometry

Of all spectroscopic techniques, mass spectrometry most nearly approaches the sensitivity of gas chromatography; it is possible to obtain adequate spectra with as little as 10^{-8} g of material. It is necessary to present the sample to the ion source of the spectrometer as a vapor, that is, in the same state as it emerges from the chromatograph. Recording of spectra can be rapid; modern instruments are capable of scanning the mass range of interest for steroid analysis in a few seconds. In these respects, then, the mass spectrometer is the ideal instrument to be coupled to the chromatograph for the direct spectral recording of fractions emerging from the column. Most gas chromatographic columns operate with their outlets at or near atmospheric pressure and it is necessary to reduce the pressure to a value suited to the ion source of the spectrometer, usually 10^{-5} Torr or less. In so doing, the majority of the sample is discarded.

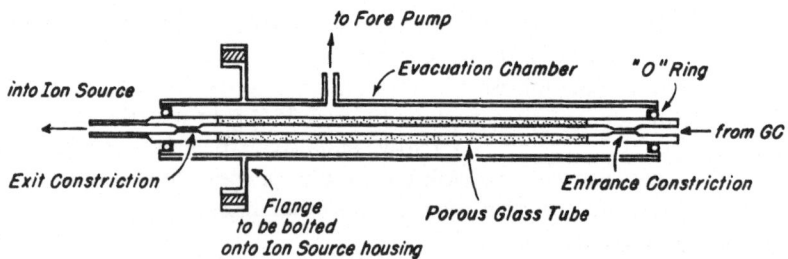

Figure 5-3. Connection of gas chromatograph to mass spectrometer. Sample enrichment device. [J. I. Watson and K. Biemann, Anal. Chem. 37:844 (1964)].

Furthermore, sensitivity is also limited by the fact that the sample issues from the column much diluted by carrier gas so that the partial pressure in the ion source is likely to be of the order of 10^{-11} Torr, which is approaching the limit of detection of even the best instruments. Recently, two methods of enrichment have been reported which take advantage of the differences in effusion properties of the sample and the carrier gas (helium). The method of Ryhage [14] uses separators which consist of suitably shaped and aligned orifices across which pressure drops are produced by differential pumping. The heavier sample molecules tend to pass straight through the orifices while the lighter helium molecules tend to diffuse into the chambers between the orifices and are pumped away. Enrichments of about one-hundred fold are claimed for the system. Biemann's method [15] relies upon the different rates of effusion of sample and carrier gas through a fritted glass tube. Details of the apparatus are shown in Figure 5-3. Enrichments of a similar order of magnitude are obtainable with this technique. Ryhage provides insufficient detail to allow an assessment of the relative merits of the two techniques. It seem likely, however, that construction of the Biemann apparatus would be simpler and it has the advantage of being made entirely of glass.

Low-resolution mass spectra of steroids have been systematically studied by a number of workers [16] so that a reasonable catalog of spectra exists. Identification by low-resolution mass spectrometry depends partially upon comparison with known spectra and also upon an analysis of the mode of fragmentation of the molecule. A considerable amount of literature exists dealing with fragmentation processes produced by electron impact; the reader is referred to the several excellent books on the subject for details [16–19].

Recently, high-resolution instruments have become available which make possible the separation of ions having the same nominal mass but differing in elemental composition and thus in fractional mass. By accurate measurement of the mass (to 5 ppm or better) it is possible to determine the elemental composition of the ion and by extension to determine the structure of the molecule [15].   The power of the technique lies in the fact that no prior assumptions are required to determine the structure of a molecule. On the other hand, it provides little or no information as to the stereochemistry of the compound studied.   The amount of information contained in a high-resolution mass spectrum is so great that manual processing is out of the question and electronic data processing techniques must be employed to deal with it. Thus, to the already high cost of the spectrometer is added the requirement that computer facilities must be available. It is unlikely that high-resolution techniques, despite their advantages, will supplant low-resolution mass spectrometry entirely.   Low-resolution instruments are now within the reach of many laboratories and it is to be expected that applications to biochemical problems will increase as the advantages of the method become more widely known.

QUANTITATIVE ANALYSIS

Accuracy

The accuracy of any gas chromatographic procedure is difficult to determine since the commonly employed detectors are not absolute. Thus any losses incurred either in the injection procedure or on the column will be approximately the same both for sample and standard. Losses from injection of sample can be largely eliminated by using the techniques described earlier.   Some loss on the column appears inevitable at the present stage of development of column packings. An estimate of the extent of column loss can be obtained from inspection of calibration curves: Ideally the curve should pass through the origin, but usually it will intersect the concentration axis at some positive value which represents the loss on the column. If possible, analytical conditions should be arranged so that column loss represents no more than 1% of the amount of compound sought.

Precision

Errors may arise from three main sources: injection of sample, the instrumentation itself and the effect of small changes in operating conditions on the column and detector, and peak measurement. In a well-designed system in good operating order it should be possible to achieve a relative standard deviation of 2–3% or better. Techniques for sample injection and the effect of system variables have been discussed earlier; in the following section methods of peak measurement are discussed.

Peak Measurement

Either peak height or peak area may be measured. Generally,

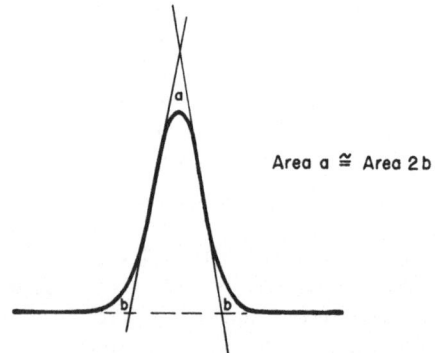

Area a $\cong$ Area 2b

Figure 5-4. Peak area measurement. Triangulation method.

measurement of peak height is more rapid and more accurate than that of peak area. However, HETP and thus peak heights and widths are often dependent upon sample size, whereas peak area is not. Thus, plots of peak area against sample size are usually linear over a wider range than those of peak height. In steroid gas chromatography, samples are normally sufficiently small to fall within the linear range of peak height plots.

Variation in carrier gas flow rate introduces errors in two ways: the effect on column efficiency and the effect on detector response. Changes in HETP, which is flow-dependent, will cause error in the measurement of peak height but not of peak area. It is usually possible to operate columns so that the minimum in the curve of HETP against flow rate is broad in order to minimize any error resulting from flow variations. Change in detector response, discussed earlier, affects both peak height and peak area measurements.

Peak area may be measured in several ways: Probably the most widely used is the method of triangulation, where tangents are drawn to the points of inflection of the peak and the triangle formed by the tangents and the baseline is measured (Figure 5-4). The method gives good results for symmetrical peaks but may introduce quite large errors if the peak is appreciably skewed (Figure 5-5). Other methods include planimetry and counting squares, both of which are quite laborious, and cutting out peaks and weighing them. This last method is quite useless since large errors may be introduced by changes in the moisture content of the paper and also because the chromatogram is destroyed.

Peak area measurement becomes tedious and time-consuming if large numbers of chromatograms must be processed. Then, it becomes worthwhile to consider the use of some form of automatic peak area integration. Two simple forms of integrator have been fairly widely employed: ball and disk, and integrating motor. Both depend upon the motion of the recorder slide-wire and integrate with respect to time all signals displacing the slide wire from its zero position. Therein lies the weakness of the system since all signals, whether they be noise, drift, or merely a constant offset of the recorder pen from zero, will be integrated in addition to the chromatographic peaks. Unless the chromatogram is virtually perfect, the amount of effort required to discard that fraction of the integral not due to the chromatographic peak is likely to be greater than that involved in measurement by triangulation. Thus, the simpler methods of integration are of little real value and cannot be recommended. Recently, more sophisticated devices have become available which can compensate for baseline drift and noise. Some method of sensing the rate of change of signal is also incorporated so that integration can be made to start at any predetermined value of slope of the chromatographic curve. Thus, any steady signal present as the result of incomplete suppression is not integrated. Also, the slope sensing device allows integration of incompletely separated peaks. Integrals are printed out on paper tape together with times for peak maxima. Such a device is relatively expensive, costing at least as much as the chromatograph itself, so that obviously its use would only be considered if manual measurement became excessively burdensome.

Measurement of Partially Separated Peaks

It is often impractical or impossible to achieve complete separation of all the peaks of interest. Two typical examples, shown in Figure 5-6, represent extreme cases where resolution is 0.75. In Figure 5-6a the two components are present in approximately equal amounts, but, although the separation is poor, the peak overlap introduces no error into the measurement of peak height. Even when there is an order of magnitude difference in the amounts of the two com-

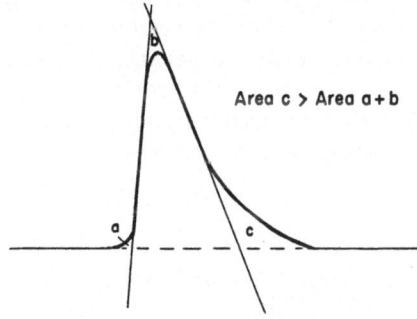

Area c > Area a + b

Figure 5-5. Peak area measurement. Error in triangulation method caused by skewed peak.

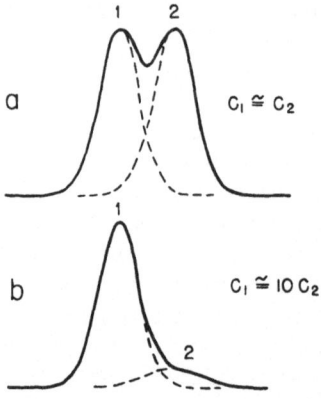

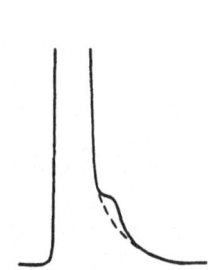

Figure 5-6. Peak height measurement   Figure 5-7. Incomplete separation of
of incompletely separated peaks.   minor component from the solvent.

ponents, it is still possible to measure the minor component with fair
accuracy. Peak area may be measured by dropping a perpendicular
from the minimum between the peaks to the baseline and measuring
areas on either side of the perpendicular. Reasonably accurate
results are obtained by this method, but values for the examples shown
in Figure 5-6b are less precise than those obtained by peak height
measurement. A third example is shown in Figure 5-7. This situation
usually arises when a minor component is incompletely separated
from the solvent peak. Reasonable results can be obtained by using
the curve of the major peak as the baseline, but precision is likely to
be of the order of ±10% or worse.

Calibration

The system may be calibrated either directly with known amounts
of the compound sought or by the internal standard method. In this
latter procedure, known amounts of a second compound, separable
from the first, are added and concentrations of the compound sought
are calculated by reference to the second standard. This technique
minimizes errors arising from the sample injected but relies on
similar behavior of both compounds. If one compound is lost preferen-
tially during the procedure, errors will result.

REFERENCES

1. M. B. Whittier and G. R. Umbreit, Facts and Methods for Scientific Research 6:3 (1965).
2. L. Dorfman, Chem. Rev. 53:47 (1953).
3. J. P. Dusza, M. Heller, and M. Bernstein, Physical Properties of the Steroid Hormones,
   edited by L. L. Engel, (Macmillan, New York, 1963).

4. C. Djerassi, Optical Rotatory Dispersion (McGraw-Hill, New York, 1960).
5. K. Dobriner, E. R. Katzenellenbogen, R. N. Jones, G. Roberts, and B. S. Gallagher, Infrared Absorption of Steroids, An Atlas, Vol. I (1953), Vol. II (1958) (Interscience, New York).
6. R. S. Juvet and S. Dal Nogare, "Gas Chromatography," in: Anal. Rev. (1964); Anal. Chem. 36:39R (1964).
7. Beckman Instruments, Inc., Model IR-102.
8. N. M. R. Abstract Service, Preston Technical Abstracts Co., 909 Pitner Avenue, Evanston, Illinois.
9. O. Jardetzky and C. D. Jardetzky, "Introduction to Magnetic Resonance Spectroscopy. Methods and Biochemical Applications," in: Methods of Biochemical Analysis, Vol. IX, edited by D. Glick (Interscience, New York, 1962).
10. J. B. Stothers, "Application of Nuclear Magnetic Resonance Spectroscopy" in: Technique of Organic Chemistry, Vol. XI, edited by A. Weissberger (Interscience, New York, 1963), Part 2.
11. J. B. Stothers, Quart. Rev. (London) 19:144 (1965).
12. N. S. Bhacca and D. H. Williams, Applications of NMR Spectroscopy in Organic Chemistry (Holden-Day, San Francisco, 1964), Vol. II.
13. E. G. Brane, Anal. Chem. 37:1183 (1965).
14. R. Ryhage, Anal. Chem. 36:759 (1964).
15. J. T. Watson and K. Biemann, Anal. Chem. 37:844 (1964).
16. H. Budzikiewicz, C. Djerassi, and D. H. Williams, Structural Elucidation of Natural Products by Mass Spectrometry (Holden-Day, San Francisco, 1964), Vol. II.
17. F. E. McLafferty, editor, Mass Spectrometry of Organic Ions (Academic Press, New York, 1963).
18. K. Biemann, Mass Spectrometry: Organic Chemical Applications (McGraw-Hill, New York, 1962).
19. J. H. Beynon, Mass Spectrometry and Its Application to Organic Chemistry (Elsevier, Amsterdam, 1960).

Chapter 6

The History of Gas–Liquid Chromatography of Steroids

The application of gas–liquid chromatography to the separation and measurement of steroids dates back less than six years. Although it appeared from the early reports of gas chromatography [1–3], that this tool would find its primary application in the separation of relatively volatile substances, by 1959 a number of compounds of boiling points nearly equal to steroid molecules had been gas chromatographed successfully. Notable among them are the fatty acids, which were chromatographed as the methyl esters to reduce the hydrogen bonding effects of the carboxyl group, thereby decreasing their boiling point. Late in 1959, there appeared a report by Eglinton et al. [4] describing the gas chromatographic separation of a number of substances including two relatively nonpolar sterols. This work was done using short grease columns and required fairly long retention times. Soon thereafter, a report by Berthuis and Recourt [5] also showed the possibility of chromatographing sterols in the vapor phase. However, the first specific efforts directed toward the separation of hormones containing the steroid nucleus were reported independently and nearly simultaneously by three laboratories in 1960. Vanden-Heuvel et al. [6] in a brief note in July, 1960, described the successful chromatography of several androstane and pregnane derivatives on thin-film columns at relatively low temperatures. At approximately the same time, Wotiz and Martin [7] reported to the American Chemical Society in September 1960, the separation, identification, and quantitative measurement of the three classical estrogens, laying the basis for a biologically useful analytical method. In October, 1960, Chen and Lantz [8] independently reported the gas chromatographic separation of steroids using both packed and capillary columns. The original report by Horning's group [6] was soon followed by several other brief reports showing a number of further separations and describing newer silicone polymer phases for steroid gas chromatography. Notable among them was the description of the

61

ketone selective phase QF-1 [9]. Lipsky and Landowne [10] further elaborated on the separation of a number of related steroids, using polyester phase rather than the silicone elastomers. The use of derivatives was specifically suggested by Wotiz and Martin [11], since this allowed the molecular-weight-based separation of the three classical estrogens on nonpolar phases. The derivatives appeared to impart greater stability to the molecule and they prevented, at least in part, the rather severe adsorption of polyhydroxy compounds on the solid support and possibly the stationary phase. Moreover, these derivatives are formed in nearly quantitative yield and are relatively stable to spontaneous hydrolysis. Further progress was made by the application of the trimethylsilyl ether derivatives, described by Langer et al. [12], to the steroids by Horning and his co-workers [13]. Despite the early difficulties encountered by some of these investigators in utilizing the highly sensitive but occasionally unstable β -ionization detectors, methodological progress continued and was considerably advanced by the development of the hydrogen flame ionization detector [14, 15] which imparted a greater range of linearity with little decrease in sensitivity. This new detector further had the advantage of cleaning itself during chromatography by burning the waste products, allowing an essentially continuous utilization of the instrument for analytical work. Soon after, there appeared a number of papers by several investigators showing specific separation of pure steroids. The first actual biological application of gas chromatographic separation and simultaneous quantitation was reported by Wotiz and Martin [16] at the Federation Meetings in March, 1961, which in essence laid the foundation for the method described later in this text. These workers were able to measure estrone, estradiol, and estriol in pregnancy urine with considerable accuracy and convenience. Data for establishing the validity of the assay were presented and were published in a later paper [17]. Simultaneously with the appearance of the second paper mentioned, a paper by Fishman and Brown [18] was published, utilizing similar procedures and comparing favorably the quantitative data obtained by GLC with that of the well-established Brown procedure [19], lending further credence to the usefulness of the more rapid gas chromatographic assay. This heralded the appearance of a number of papers from several laboratories which had become interested in the application of this tool to biological assay procedures. Although, as pointed out before, the application of this tool to biochemical and medical research is less than six years old, methods for the determination of a large variety of steroid hormones are presently recorded in the literature and in some instances, techniques for the estimation of steroids not previously possible by other means have been reported (pregnanolone, cortol, cortolone, and certain estrogens). Some of the more common types of hormone assays can now be done more quickly and conveniently by GLC than by previously reported procedures. Indeed, the time factor alone may have great bearing on future decisions about the choice of methods to be used in individual laboratories.

Specific methods for the determination of 17-ketosteroids have been described by Cooper and Creech [20], Hartman and Wotiz [21], Sparagana et al. [22], VandenHeuvel et al. [23], Kirschner and Lipsett [24, 25], France et al. [26], and Thomas [27]. Cooper and Creech, and Hartman and Wotiz, essentially chromatographed a crude neutral urine extract after derivative formation, letting the chromatographic column do all the separating, but differing in the choice of stationary phase. Kirschner and Lipsett, on the other hand, utilizing essentially the same procedure described by Hartman and Wotiz, do include a preliminary separation using thin-layer chromatography. The purpose of the latter appears to be primarily a separation of the pregnanediol, pregnanetriol, and mixed 17-ketosteroid fractions. In contrast, Sparagana et al. carry through a most elegant, though involved, preliminary purification procedure based largely on the more classical column chromatographic separations. Gas chromatography is used primarily as a final separating tool as well as a sensitive analytical measurement.

Measurements of pregnanediol have been reported by several investigators. In 1962, Cooper et al. [28] described the direct measurement of free pregnanediol in pregnancy urine. Wotiz [29, 30] reported the direct measurement of urinary pregnanediol as the diacetate at levels below that found in the male and the nonpregnant female. As mentioned before, Kirschner and Lipsett [24] were able to measure pregnanediol as the diacetate only. Turner and co-workers [31] reported a method of pregnanediol measurement using the free steroid extracted from pregnancy urine, as did Patti et al. [32]. More recently, Jansen [33] and Cox [34] also reported preliminary details of a method of analysis of pregnanediol using gas–liquid chromatography. The measurement of plasma progesterone, a much more difficult procedure, has been described by Yannone et al. [35], by Collins and Somerville [36] using TLC and gas–liquid radiochromatography, and quite recently by VanderMolen et al. [37].

The rapid gas chromatographic determination of the acetates of the three classical estrogens in pregnancy urine has been reported by Wotiz and Martin [17, 38] and Fishman and Brown [18] and has been treated in a somewhat different manner by Yousem [39] and Touchstone [40]. Luukkainen et al. [41] also reported the use of GLC for estrogen measurements, as did Luukkainen and Adlerkreutz [42]. A method for the assay of estrogens in plasma has been reported by Kroman and co-workers [43]. Despite considerable efforts, in this laboratory as well as in a number of others, the measurement of urinary estrogens in normal menstruating and postmenopausal females has not been successfully accomplished by the use of gas chromatography alone. Wotiz and Chattoraj [44, 45] recently presented a versatile method allowing either the separation of the three classical estrogens, using one thin-layer chromatogram, or the separation of seven estrogen metabolites in human urine, requiring the use of two thin-layer chromatograms. In each case final separation and measurement are achieved through GLC following derivative formation.

Measurement of testosterone in the urine has been described by several investigators. The work of Futterweit [46] describes the measurement of this hormone involving preparation of a ketonic fraction with Girard T reagent, followed by TLC and gas–liquid chromatography. Several papers describing somewhat more direct methods appeared recently [47–50].

In 1963, Guerra-Garcia et al. [51] were able to utilize gas chromatography to measure the concentration of testosterone in plasma from men. However, the sensitivity of this procedure was insufficient for the measurement of testosterone in plasma from women. The use of an electron capture detector following the formation of proper derivatives of testosterone was described by Brownie et al. [52]. These workers have utilized the steroid mono-chloroacetates, as described by Landowne and Lipsky [53]. A possible application of the more volatile and more electron-attractive heptafluorobutyrates described earlier by Clark and Wotiz [54] is presently under investigation in our laboratory.

The determination of pregnanetriol and certain steroids closely related to it has been described by Rosenfeld et al. [55] and includes a prior separation by paper chromatography followed by gas chromatography. Pregnanetriol itself can be analyzed more readily by the method of Kirschner and Lipsett [24], where this substance appears as an individual fraction in the thin-layer chromatogram. A direct assay in a mixture containing pregnanediol and the 17-ketosteroids was alluded to by Creech [56], but no biological data are as yet available.

The analysis of urinary corticosteroids is a much more complex picture and rather than discuss it in any detail at this point, let it suffice to say that a number of methods for the measurement of certain specific glucocorticoids have been described. Nearly all of them involve degradation of the side chain by periodic acid [57], bismuthate [58, 59], or pyrolysis [60]. The analysis of certain glucocorticoids without side-chain elimination was described by Kirschner and Fales [61], using the bismethylene dioxides, and by

Rosenfeld, using the TMSi derivative [62]. A very elegant screening procedure for ketosteroids and ketogenic steroids derived from the corticosteroids has been described by Menini and Norymberski [63].

Although aldosterone, the principal mineralocorticoid in man, has been gas chromatographed [64], no method of assay of this compound in urine or plasma has been developed as yet. On the other hand, Carr et al. [65] did describe a method for the assay of urinary tetrahydroaldosterone, one of the prime metabolites of aldosterone.

A number of other procedures have been reported in the literature which might well be of specific interest to one investigator or another. The isolation of fecal steroids has been described by Eneroth et al. [66] and Wells and Makita [67], and several methods for the measurement of blood sterols are also available [68–70]. A few investigators have utilized the great speed and high sensitivity of GLC in metabolic studies involving both *in vivo* and *in vitro* experiments. In 1961, Gower and Haslewood [71] showed the formation of Δ^{16}-androsten-3a-ol from acetate by testis slices and in 1962, Knights et al. [72] isolated 5a-pregnane-3a, 6a, 20a-triol following administration of progesterone to rabbits. In 1963, Kirschner et al. [73] applied GLC methods to a study of the metabolism of DHEA in man and in 1964, Thomas [74] reported on a study of the metabolism of Δ^4-androstene-3, 17-dione in guinea pigs.

Another interesting aspect of this relatively new tool in the armamentarium of the steroid chemist has been the study relating relative retentions of steroids to their structure. Several interesting papers showing not only the effect of different functional groups on retention times (ΔRM_G) but also describing simple test-tube reactions (ΔRM_r) to allow characterization of specific compounds, are available in the literature. A more detailed discussion, references, and possible applications of this information are presented as a separate chapter.

REFERENCES

1. A. T. James and A. J. P. Martin, Analyst 77:915 (1952).
2. A. T. James and A. J. P. Martin, Biochem. J. 50:679 (1952).
3. N. H. Ray, J. Appl. Chem. 4:21, 82 (1954).
4. G. Eglinton, R. J. Hamilton, R. Hodges, and R. A. Raphael, Chem. Ind. 955 (1959).
5. R. K. Berthuis and J. H. Recourt, Nature 186:372 (1960).
6. W. J. A. VandenHeuvel, C. C. Sweeley, and E. C. Horning, J. Am. Chem. Soc. 82:3481 (1960).
7. H. H. Wotiz and H. F. Martin, Am. Chem. Soc. Meeting, New York City, September 1960.
8. C. Chen and C. D. Lantz, Biochem. Biophys. Res. Commun. 3:182 (1960).
9. W. J. A. VandenHeuvel, E. O. A. Haahti, and E. C. Horning, J. Am. Chem. Soc. 83:1513 (1961).
10. S. R. Lipsky and R. A. Landowne, Anal. Chem. 33:818 (1961).
11. H. H. Wotiz and H. F. Martin, J. Biol. Chem. 236:1312 (1961).
12. S. H. Langer, R. A. Friedal, J. Wender, and A. C. Sharkey, Anal. Chem. 30:1353 (1958).
13. T. Luukkainen, W. J. A. VandenHeuvel, E. O. A. Haahti, and E. C. Horning, Biochim. Biophys. Acta 52:599 (1961).

14. J. Harley and V. Pretorius, Nature 181:177 (1958).
15. I. G. McWilliam and A. Dewar, Nature 182:1664 (1958).
16. H. H. Wotiz and H. F. Martin, Federation Proc. 20:199 (1961).
17. H. H. Wotiz and H. F. Martin, Anal. Biochem. 3:97 (1962).
18. J. Fishman and J. B. Brown, J. Chromatog. 8:21 (1962).
19. J. B. Brown, Biochem. J. 60:185 (1955).
20. J. A. Cooper and B. G. Creech, Anal. Biochem. 2:502 (1961).
21. I. S. Hartman and H. H. Wotiz, Steroids 1:33 (1963).
22. M. Sparagana, W. B. Mason, and E. H. Keutman, Anal. Chem. 34:1157 (1962).
23. W. J. A. VandenHeuvel, B. G. Creech, and E. C. Horning, Anal. Biochem. 4:191 (1962).
24. M. A. Kirschner and M. B. Lipsett, J. Clin. Endocrinol. Metab. 23:255 (1963).
25. M. A. Kirschner and M. B. Lipsett, Steroids 3:277 (1964).
26. J. T. France, N. L. McNiven, and R. I. Dorfman, Acta Endocrinol. Suppl. 90:71 (1964).
27. B. S. Thomas, Airlie House Conference, Warrenton, Va., February 1965.
28. J. A. Cooper, J. P. Abbott, B. K. Rosengreen, and W. R. Claggett, Am. J. Clin. Pathol.
    38:388 (1962).
29. H. H. Wotiz, Biochim. Biophys. Acta 69:415 (1963).
30. H. H. Wotiz and P. J. Mozden, Human Ovulation, edited by C. S. Keefer (Little, Brown and
    Co., Boston, 1965), p. 160.
31. D. A. Tutner, G. E. Seegar-Jones, I. J. Sarlos. A. C. Barnes, and R. Cohen, Anal. Biochem.
    5:99 (1963).
32. A. A. Patti, P. Bonnano, A. A. Stein, and T. Frawley, Am. J. Clin. Pathol. 39:399 (1963).
33. A. P. Jansen, Clin. Chim. Acta 8:785 (1963).
34. R. I. Cox, J. Chromatog. 12:242 (1963).
35. M. E. Yannone, D. B. McComas, and A. Goldfien, J. Gas Chromatog. 2:30 (1964).
36. W. P. Collins and I. F. Somerville, Nature 203:836 (1964).
37. H. VanderMolen and D. Groen, Airlie House Conference, Warrenton, Va., February 1965.
38. H. H. Wotiz, Biochim. Biophys. Acta 74:122 (1963).
39. H. L. Yousem, Am. J. Obstet. Gynecol. 88:375 (1964).
40. J. C. Touchstone, A. Nikolski, and T. Murawec, Steroids 3:569 (1964).
41. T. Luukkainen, W. J. A. VandenHeuvel, and E. C. Horning, Biochim. Biophys. Acta 62:153
    (1962).
42. T. Luukkainen and H. Adlerkreutz, Biochim. Biophys. Acta 70:700 (1963).
43. H. S. Kroman, S. R. Bender, and R. L. Capizzi, Clin. Chim. Acta 9:73 (1964).
44. H. H. Wotiz and S. C. Chattoraj, Anal. Chem. 36:1466 (1964).
45. H. H. Wotiz and S. C. Chattoraj, Airlie House Conference, Warrenton, Va., February
    1965.
46. W. Futterweit, N. L. McNiven, L. Marcus, C. Lantos, M. Drosdowsky, and R. I. Dorfman,
    Steroids 1:628 (1963).
47. H. Ibayashi, M. Nakamura, S. Murakawa, T. Uchikawa, T. Tanioka, and K. Nakao, Steroids
    3:559 (1964).
48. R. V. Brooks, Steroids 4:117 (1964).
49. D. H. Sandberg, N. Ahmad, W. W. Cleveland, and K. Savard, Steroids 4:557 (1964).
50. M. Sparagana, Steroids 5:773 (1965).
51. R. Guerra-Garcia, S. C. Chattoraj, L. J. Gabrilove, and H. H. Wotiz, Steroids 2:605 (1963).
52. A. C. Brownie, H. J. VanderMolen, E. E. Nishizawa, and K. B. Eik-Nes, J. Clin. Endocrinol.
    Metab. 24:1091 (1964).
53. R. A. Landowne and S. R. Lipsky, Anal. Chem. 35:532 (1963).
54. S. J. Clark and H. H. Wotiz, Steroids 2:535 (1963).
55. R. S. Rosenfeld, M. C. Lebeau, R. D. Jandorek, and T. Sabina, J. Chromatog. 8:355 (1962).
56. B. G. Creech, J. Gas Chromatog. 2:194 (1964).
57. G. W. Kittinger, Steroids 3:21 (1964).
58. E. Bailey, J. Endocrinol. 28:131 (1964).
59. E. Bailey, Airlie House Conference, Warrenton, Va., February 1965.
60. J. A. Luetscher and G. R. Gould, J. Chromatog. 13:350 (1964).
61. M. A. Kirschner and H. M. Fales, Anal. Chem. 34:1548 (1962).
62. R. S. Rosenfeld, Steroids 4:147 (1964).
63. E. Menini and J. E. Norymberski, Biochim. J. 95:1 (1965).
64. H. H. Wotiz, I. Naukkarinen, and H. E. Carr, Biochim. Biophys. Acta 53:449 (1961).
65. H. E. Carr and H. H. Wotiz, Biochim. Biophys. Acta 71:178 (1963).
66. P. Eneroth, K. Hellstrum, and R. Ryhage, J. Lipid Res. 5:245 (1964).

67. W. Wells and M. Makita, Anal. Biochem. 4:204 (1962).
68. H. J. O'Neill and L. L. Gershlein, Anal. Chem. 33:182 (1961).
69. L. P. Cawley, B. O. Musser, S. Campbell, and W. Faucette, Am. J. Clin. Pathol. 39:450 (1963).
70. C. J. W. Brooks and J. S. Young, Biochem. J. 84:53P (1962).
71. D. B. Gower and G. A. D. Haslewood, J. Endocrinol. 23:253 (1961).
72. B. A. Knights, A. W. Rogers, and G. H. Thomas, Biochem. Biophys. Res. Commun. 8:253 (1962).
73. M. A. Kirschner, M. B. Lipsett, and H. Wilson, Acta Endocrinol. 43:387 (1963).
74. G. H. Thomas, E. Forchielli, and K. Brown-Grant, Nature 202:260 (1964).

Chapter 7

Column Practice

The simplified column theory presented earlier must be regarded primarily as a qualitative guide to the achievement of column efficiency. The production of good analytical columns depends almost entirely upon the skill of the operator and thus it is of the utmost importance that the newcomer to gas chromatography acquire proficiency in making columns.

Manufacturers of chromatographs and suppliers of column materials have devoted much effort in recent years to the production of efficient columns for steroid analysis, and it is now possible to buy columns suitable for most work. However, the occasion always arises where the commercial product is either not good enough or unsuitable for the application and then it becomes necessary for the experimenter to prepare his own columns.

TUBING MATERIAL

Glass and stainless steel tubing are the most commonly used materials for the construction of high-temperature columns. The relative merits of these two materials for steroid chromatography has been the subject of some controversy. It seems certain that there is little to choose between them with respect to column efficiencies and that most of the loss of sample ascribed to the use of metal columns is in fact a function of the method of injection of the sample.

Metal columns have the advantages of durability and ease of fabrication; on the other hand, the condition of the packing in a glass column is readily observed. Certain compounds, particularly those containing halogen, may react with hot metal surfaces; then, of course, it is essential to use glass columns.

COLUMN CONFIGURATION

Theoretically, a straight column would be expected to be more efficient than a coiled one since the difference in path lengths between the inside and outside of the coil contributes to peak spreading. In practice, the effect is small provided that the radius of curvature of the coil is large compared with the column bore.

Metal columns coiled before packing will often have higher efficiencies than those packed before coiling. The tubing is deformed by the coiling process and the packing is crushed to some extent, leading to a decrease in column efficiency.

COLUMN DIAMETER

Within the range normally used for analytical gas chromatography (2–10 mm), diameter has little effect on efficiency. Reports at various times have favored one column diameter over another, but it seems likely that differences in performance can be ascribed to causes other than change in column diameter.

SOLID SUPPORT

The ideal solid support should have large surface area, uniform pore size, and should be completely inert.

Pore Size

In practice, only the first of these requirements is met by materials at present available to the chromatographer. Pore size of commercial materials varies over a rather wide range and this variation has the effect of causing local variations in the thickness of the film of liquid phase. All that the user can do is to ensure that the gross distribution of liquid phase is as uniform as possible.

Adsorptivity of Support

Much more serious than nonuniformity of pore size is surface activity of the support. Steroids are polar molecules which will interact readily with any adsorptive surface and, moreover, the stationary phase film thickness normally used in steroid chromatography is small enough so that the support may be incompletely covered. Thus, support surface activity is a critical factor in the success or failure of the technique. Adsorption by the support manifests itself in two ways: as loss of sample in the column and as distortion, particularly tailing, of the chromatographic peak.

Support effects may be minimized in several ways: by formation of derivatives that reduce the polarity of the steroid molecule, by employing stationary phases that interact with the support to a much greater extent than does the sample, by treatment of the support to reduce its surface activity. The use of derivatives has advantages apart from any effect on support adsorption and the choice of stationary phase may rest with considerations other than support effects. In any case, these two approaches are only palliative and it is more

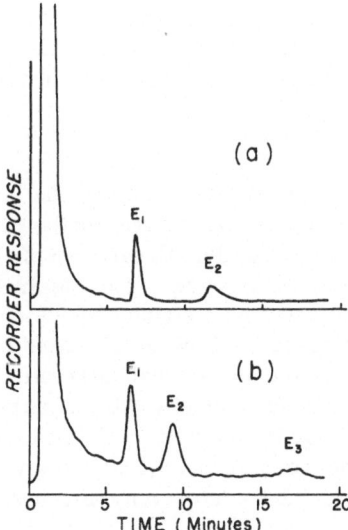

Figure 7-1. Sorption by support material. (a) Commercial acid-washed, silanized support. (b) Same support after exhaustive silanization. Column: 4 ft × 2 mm glass, 3% SE-30, 220°C. 45 ml/min N_2. Sample: 0.025 μg each of the acetates of estrone, estradiol, and estriol.

satisfactory to improve the characteristics of the support itself. Figure 7-1a illustrates the effect of adsorption upon the chromatogram and Figure 7-1b shows the results obtained under identical conditions after further treatment of the support. The original support was a commercial material which had been treated by the manufacturer; the improved support was the same material after further extensive treatment in the laboratory. For both materials, treatment consisted of washing with acid followed by silanization; the difference between them is due only to the extent of the treatment. Until quite recently even the better commercial supports produced results similar to those in Figure 7-1a, but during the last few months at least one manufacturer [1] has announced the availability of specially treated material which is as least as good as anything we have been able to produce in the laboratory. The material is expensive but the price is fully justified by the results obtained.

Particle Size

Theory predicts, and practice confirms, that particle size affects column efficiency. Within limits, higher column efficiencies are attained with smaller particles. Particle size in the ranges 80–100 or 100–120 mesh are generally suitable. There is no reason, in principle, why smaller particle sizes should not be used. However, the very high pressures required to obtain optimum flow in columns using, say, 400 mesh particles, introduce practical difficulties.

Mechanical Strength

The diatomaceous earth from which the support material is prepared is rather fragile so that attrition of the particles by vibration or washing occurs rather easily. Breakdown of the particles results in a range of particle size wider than desirable and produces an excessive amount of very fine particles which contribute disproportionately to the resistance to flow of a packed column. In addition, fracture of a particle exposes a fresh untreated surface so that any treatment to reduce surface activity will be partially nullified. Recently, a new support has been introduced that is considerably denser and much less prone to fracture [2]. However, this material, as supplied, exhibits too much adsorptivity for use in steroid chromatography.

It is obvious from the above that care must be taken when handling the support to minimize particle breakdown. In general, sieving of support material is not recommended. The material, as received from the manufacturers, normally falls within the specified mesh range except for the presence of fines. We have found that washing with methanol removes most of the fines by flotation. Methanol is used rather than water for two reasons: It is more easily removed in the subsequent drying process and any chlorine groups that may remain as a result of the silanization process are removed.

Further operations involving the support, i.e., coating or column packing, should also be conducted so that mechanical handling is minimized.

STATIONARY PHASE
Volatility

The life of a gas chromatographic column is determined primarily by the rate of loss of stationary phase by volatilization. Thus, the first prerequisite of a stationary phase is that its vapor pressure be low at the required operating temperatures. For steroids, it is usually necessary to operate in the range $200-250°$ and thus the choice of stationary phase is limited by the temperature requirement.

The effect of stationary phase bleed upon the detector must also be considered. The ionization detectors are, of course, as sensitive to volatile stationary phase as they are to sample, and problems of detector noise and contamination are largely associated with bleed from the column. The extent to which the problem occurs depends upon the particular stationary phase or detector employed. For example, use of a QF-1 stationary phase with a flame detector is satisfactory since the flame is comparatively insensitive to halogenated material. On the other hand, satisfactory use of this stationary phase with an electron capture detector is difficult since the detector is highly sensitive to halogenated compounds.

Stability

A second property affecting rate of loss of stationary phase is chemical stability. Unfortunately, the stability of high-temperature phases leaves something to be desired since both volatility and stability tend to decrease with increasing molecular weight. All the stationary phases now in use for steroid separations are polymeric materials; these are sufficiently thermally unstable so that operation above approximately 280° is impractical. Interaction with the support, or impurities in the carrier gas, may contribute to stationary phase breakdown. Catalytic action of the support can be traced to the presence of metallic impurities or to traces of acid remaining after pretreatment. Traces of oxygen in the carrier gas may cause oxidative degradation of the stationary phase, particularly if polyester phases are employed. Thus, care in the choice of support and in ensuring that the carrier gas is as pure as possible can assist in prolonging column life.

Selectivity

Stationary phases may be classified in two broad categories: nonspecific, separating primarily by differences in boiling point, and selective. Selectivity of a stationary phase can be specified only with reference to sample type, since a particular phase may function in both categories depending upon the type of sample. It should also be realized that a phase that is selective with respect to a particular class of compound will probably function as a boiling-point separator for comparison within that class. The relationship between steroid structure and retention time is discussed in detail in a later chapter. The relationship is complex and prediction of the relation of compounds not previously chromatographed can only be qualitative at best. However, enough information is now available to allow identification of unknown compounds from their chromatographic behavior.

COLUMN PREPARATION
Coating of Support

It is worth re-emphasizing that the application of a uniform coating of stationary phase is essential for the production of a good column.   Of all the techniques recommended in the literature, the solution coating technique [3] has given us the most consistently good results. The detailed procedure is as follows:

Pour 15–20 g of the appropriate support material (e.g., Gas Chrom Q) into a beaker and add 50–100 ml of dry methanol. Swirl gently, allow to settle for 30 sec., and pour off the supernatant liquid; repeat the flotation procedure until the washing solution is free from fines.   Transfer the washed support to an evaporating dish and dry at 110°C for 2–3 hr. Pour the dried support into a fritted glass Buchner funnel and add 100 ml of solution of the stationary phase of concentration equal to the required concentration of stationary phase on the support (e.g., for a 3% coating use a 3% w/v solution). Swirl gently so as to wet the support thoroughly and remove excess solution by gentle suction.   Transfer the coated material to an evaporating dish and dry at 110°C until the powder is free flowing.

Mechanical handling of the support material must be as gentle as possible at all times in order to minimize attrition. As long as a severalfold excess of the stationary phase solution is used, the amount of stationary phase deposited on the support will be equal to the concentration in solution and independent of the volume of solution used.

Column Packing

Straight Columns.   Insert a small plug of glass wool into one end of the tube and attach a funnel to the opposite end by means of a short length of rubber tubing. Pour a small amount of packing material (enough to fill 2–3 in. of column) into the funnel and tap the column gently to consolidate the packing. Repeat the process until all but about $1/4$ in. of the column is filled, remove the funnel, and plug the open end with glass wool.

U-Shaped Columns.   Attach a funnel to one arm of the U and follow the procedure given in the previous paragraph. When the first arm is full, transfer the funnel to the second arm and repeat. Ensure that both arms are filled to within $1/4$ in. of the end of the tubes and plug with glass wool.

Coiled Columns.   Plug one end of the column with glass wool and attach the same end to a water vacuum pump. Attach a funnel to the opposite end and fill the column as described in the section "Straight Columns," applying gentle suction.   Manipulate the column so as to prevent small amounts of packing collecting at the lowest points of otherwise unfilled coils. Once the packing operation is completed, plug the open end of the column with glass wool.

Pyrex glass wool should be used for plugging columns; it is sometimes found that the glass wool is excessively adsorptive, necessitating treatment before use according to the following procedure:

Place a few grams of glass wool (Pyrex No. 7220) in a round-bottomed flask, add an excess of a 10% solution of hexamethyldisilazane in dry isooctane, and boil gently under reflux for 48 hr. Pour off the reagent solution and dry the glass wool at 110°C. Store in a tightly stoppered bottle.

Column Conditioning

Newly prepared columns invariably exhibit high bleed rates owing to the presence of traces of solvent and of low-molecular-weight fractions of the stationary phase. Thus, it is necessary to maintain the column at or near its maximum operating temperature until the bleed rate has stabilized at a low value. Generally, a period of 24–48 hr must elapse before the column is suitable for analytical use.

REFERENCES

1. Applied Science Laboratories, P. O. Box 140, State College, Pa., Gas Chrom Q.
2. Johns-Manville Products Company, Celite Division, 22 E. 40th Street, New York, Chromosorb G.
3. E. C. Horning, E. A. Moscatelli, and C. C. Sweeley, Chem. Ind. (London) 751 (1959).

Chapter 8

Decomposition of Steroids During GLC

The question of the stability of steroid molecules undergoing separation on a gas chromatographic column is of course of the utmost importance. Although it is entirely feasible to measure steroids as their decomposition products rather than in their original state, it is generally more desirable to do the latter. One of the prime reasons for trying to avoid decomposition during chromatography is the rather hazardous quantitative aspect of such procedures [1]. This is not meant to imply that alteration of compounds, such as side-chain cleavage or cyclization prior to chromatography, would be necessarily undesirable.

Since in our own laboratory earliest interest was centered on the estrogens, estradiol as the representative of this group, was examined with respect to its thermal stability under a variety of conditions. Martin and Wotiz [2] described the stepwise thermal breakdown of estradiol-17β in a high vacuum (0.02 Torr) over a range of temperatures from 275° – 375°C. The work was carried out in a micro-isoteniscope requiring fairly high concentrations of material to be examined. On the average, 50 mg or more were used for each experiment. It should be pointed out that even at 0.02 Torr there still remains a significant concentration of oxygen in the apparatus. Despite this it was found that the rate of decomposition between 275° and 317° was less than 0.02% per min. The primary reaction which was observed was the elimination of hydrogen from position 17 with the formation of estrone. Significant decomposition did not occur until a temperature of 375° was reached, and even at that point less than 0.5% of the total material was altered. Following these early experiments the acetates of the three classical estrogens were chromatographed in $\frac{1}{4}$-in. copper columns, packed with 30% stopcock grease on diatomaceous earth, and with the use of a stainless steel flash heater. The material in the effluent stream (approximately 100–200 μg per injection) was trapped in liquid nitrogen and subjected to several tests, most significant among them being of course the comparison of the infrared spectra before and after chromatography. Figures 8-1, 8-2,

77

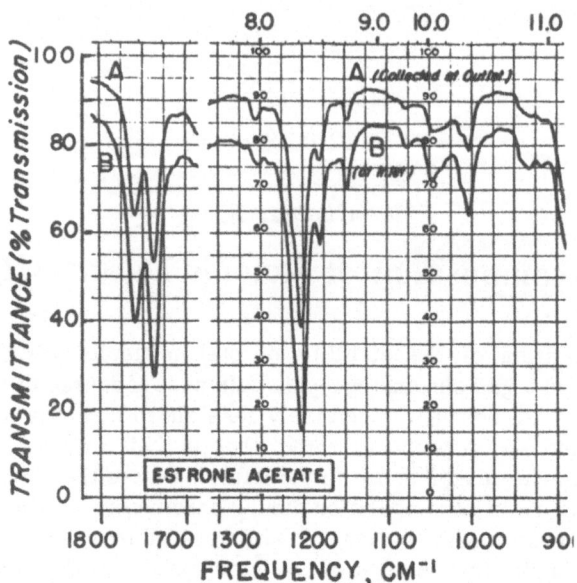

Figure 8-1. Comparison of the infrared spectra of estrone acetate before and after injection into the gas chromatograph [H.H. Wotiz and H.F. Martin, J. Biol. Chem. 236: 1312 (1961)].

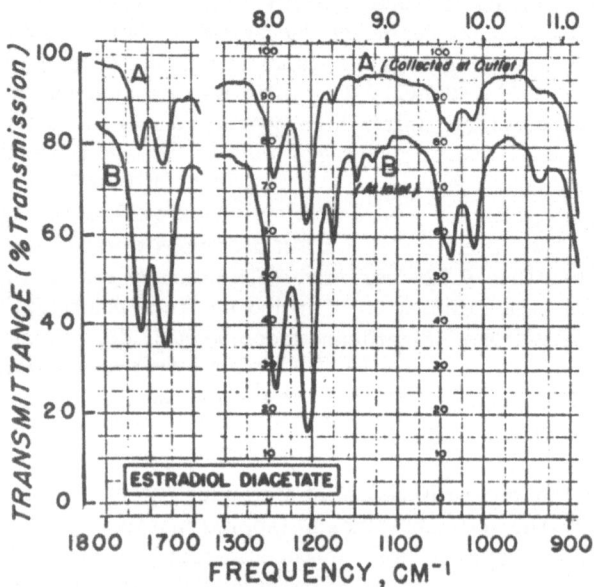

Figure 8-2. Comparison of the infrared spectra of estradiol diacetate before and after injection into the gas chromatograph [H.H. Wotiz and H.F. Martin, J. Biol. Chem. 236:1312 (1961)].

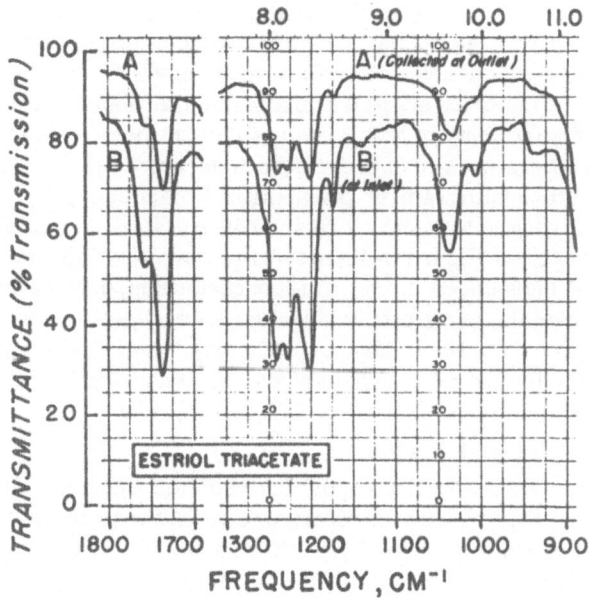

Figure 8-3. Comparison of the infrared spectra of estriol triacetate before and after injection into the gas chromatograph [H.H. Wotiz and H.F. Martin J. Biol. Chem. 236:1312 (1961)].

and 8-3 give ample evidence that under the conditions of the chromatogram no decomposition had occurred. Following the important contribution by VandenHeuvel et al. [3] describing the use of high-resolution, low-concentration columns for steroid fractionation, we noted occasionally that an overaged column or one that had been exposed improperly to high temperatures would present chromatographic records highly indicative of thermal decomposition of the materials passed through it. The need to investigate this particular aspect of gas chromatography was further enhanced by the rather vague allusions made by some investigators regarding the use of metal columns. Since in our experience copper, glass, and stainless steel columns appeared to have worked very well and to have caused no difficulties, we undertook to examine this particular problem briefly. Basically, one must examine three different areas in the instrument and the column, any one of which might be the cause of decomposition. These are the flash heater, the column container, and the column packing.

EFFECT OF FLASH HEATER

The first exposure of the steroid to heat is in the vaporizer, which in general is kept 20° to 30° above the temperature of the column itself. It soon became apparent that the improper design of such flash heaters could be a serious contributor to decomposition, particularly those heaters which have too small a mass and allow the existence of hot spots. The insertion of a glass tube inside the flash heater, allowing better heat distribution and avoiding direct metal contact at these elevated temperatures, can prevent such decomposition. Figure 8-4a shows the chromatogram of pregnanediol diacetate and cholesterol acetate following injection in a flash heater designed to accommodate such a glass tube. Figure 8-4b, in contrast, shows a chromatogram of the identical substances in the same concentration, in the same instrument, and over the same column, except that the glass lining had been removed from the flash heater.   Clearly then, this is one area of difficulty which must be avoided and can be controlled.

EFFECT OF COLUMN PACKING

The second area of decomposition remains with the column. The problem here may lie with the container or with the packing. Table 8-1 shows some results obtained with different polar and nonpolar stationary phases for a variety of steroids from the thermally rather stable estrogens to the relatively unstable corticosteroids. It can be seen quite readily that when sufficiently large concentrations (3%) of the nonpolar stationary phase (SE-30) are used, all of the substances tested can be chromatographed as single peaks provided derivatives are formed.   As the concentration of the stationary phase decreases significantly, evidence for decomposition begins to appear. As the

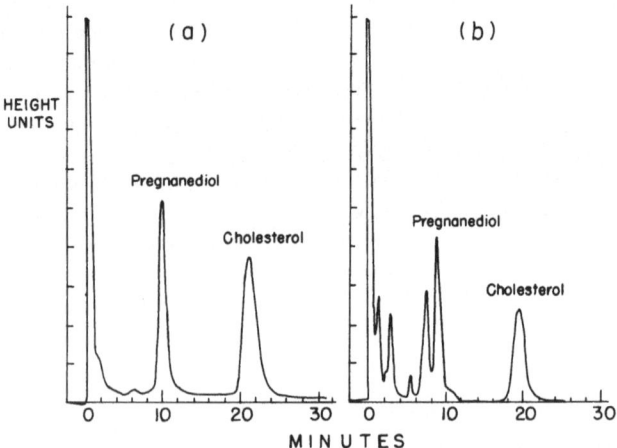

Figure 8-4. Chromatogram of a mixture of pregnanediol diacetate and cholesterol acetate on a 3% SE-30 (6 ft × ¼ in.) column at 250°C, 60 ml N_2 per min. The vaporizer was kept at 300°C. (a) With glass insert in stainless steel injector. (b) Glass insert removed.

TABLE 8-1. Effect of Stationary Phase Concentration on
Decomposition (D = decomposition, I = chromatographed intact)*

Steroid acetates†	SE-30					EGIP ‡		
	0.25%	0.5%	1%	1.5%	3%	0.5%	1%	1.5%
Estrone	D	I	I	I	I	D	I	I
Estradiol	D	I	I	I	I	D	I	I
Estriol	D	D	I	I	I	D	D	I
Pregnanediol	D	D	I	I	I	D		
Cortisone	D	D		D	I	D	D	I
Aldosterone	D	D		D	I	D	D	D

*Reproduced by permission from H.H. Wotiz, Biochim. Biophys. Acta 63:180–185 (1962),
Table I. Based on appearance of single symmetrical peak.
†All unhindered hydroxyl groups acetylated in pyridine–acetic anhydride.
‡Ethylene glycol isophthalate.

amount of stationary phase decreases, aldosterone and cortisone exhibit signs of decomposition, while estrone and estradiol may be
chromatographed with impunity at concentrations of as low as 0.5%
stationary phase.

That this decomposition is temperature-independent can be seen
from Table 8-2, which shows that aldosterone diacetate, the most heat
labile of the substances investigated, was chromatographed successfully
between 195° and 243°. Indeed, this substance has been chromatographed successfully (as a single peak) at a column temperature of
280°C. Evidence for the structural retention of the substance after
chromatography was obtained from the infrared spectrum. A later
report on this same subject by Kliman [4] suggested these findings
to be partly in error. Following acetylation with radioactive acetic
anhydride and gas chromatography, Kliman found that aldosterone retained only one of the two acetyl groups and that estradiol and estriol
each lacked one of the two or three acetyl functions, respectively. It
should be pointed out that these experiments were significantly different from those reported by us. First of all, chromatography was
carried out on a 2% column which may be deleterious to the molecule.
Examination of Table 8-1 shows that this lies at the borderline of
structural retention for aldosterone diacetate. This explanation,
however, does not allow for the differences found for the estrogens,
since all three are known to chromatograph intact on columns with
much lower concentrations of stationary phase. An explanation can be

TABLE 8-2.   Effect of Temperature on Decomposition
(D = decomposition, I = chromatographed intact)*

Column	Temperature (°C)	Steroids†				
		Aldosterone diacetate	Cortisone acetate	Estrone acetate	Estradiol diacetate	Estriol triacetate
SE-30 3% 3 ft. by ¼ in.	195	I	I	I	I	I
	205	I	I	I	I	I
	210	I	I	I	I	I
	223	I	I	I	I	I
	237	I	I	I	I	I
	243	I	I	I	I	I
EGIP ‡ 1.5% 4 ft. by ³/₁₆ in.	240	D	D			
	232	D	D			
	224	D	D			
	212	D	D			
	198	D	D			

*Reproduced by permission from H.H. Wotiz, Biochim. Biophys. Acta 63:180-185 (1962), Table II. Based on appearance of single symmetrical peak.
†All unhindered hydroxyl groups acetylated in pyridine–acetic anhydride.
‡Ethylene glycol isophthalate.

obtained from the second significant difference from our work. Because of the use of radioactive acetic anhydride, Kliman worked with amounts far below that found necessary by us for complete acetylation. If the ratio of acetic anhydride to pyridine falls significantly below one, the C-17 hydroxyl of estradiol or estriol is not readily acetylated. Indeed, we found that rapid conversion in quantitative yield occurs only when the acetic anhydride/pyridine ratio approaches five.

The evidence appears to be fairly conclusive that the lower limit of stationary phase concentration can become critical. In general, we do not recommend columns with less than 2% of stationary phase.

EFFECT OF COLUMN CONTAINER

The third factor relates to the nature of the column container. To date, no evidence has been presented that decomposition of steroids occurs because of the presence of metal columns. Whittier et al. [5] reported that they were unable to find any quantitative difference in the detectable cholesterol and cholestane when comparing stainless steel and glass columns. The only apparent difference may be a slightly greater tailing of the peaks passing through the metal column. The statistical significance of this tailing effect has not been made available. These authors came to the conclusion that in columns with less than 2% stationary phase too many exposed areas of the solid support prevent adequate passage of the steroids through the column, a finding analogous to that described earlier in this chapter. Fales, in a recent communication [6], produced more detailed evidence on this subject, verifying the usefulness of metal columns.

Despite the avoidance of the three major causes of steroid decomposition (flash heater, low concentration of stationary phase, and column material), there are presently recorded in the literature a few steroids which are sufficiently heat-sensitive to prevent their chromatography without some alteration of the molecule, at least within the limits of the experiments described. The first instance refers to C-21 steroids containing an α-ketol side chain with the hydroxyl group in the 17-position. It was shown by VandenHeuvel and Horning [7] that when such steroids are chromatographed even in an all glass system, side-chain cleavage occurs. Brooks [8] also showed the D-homo analog formation of a series of corticosteroids acetylated at C-21 only. Whether this internal rearrangement could be avoided by using a sufficiently high concentration of stationary phase cannot be ascertained at this time, since this investigator used only a 2% coating in his work. This latter view is enhanced by the observation that more extensive degradation occurs on aged columns. The authors suggest that this is caused by the accumulation of deposits at the top

TABLE 8-3. Gas Chromatography of Some Androstanes[*]

Compound	$r^{\dagger}$
5α-Androstan-17β-ol	0.42
Δ^4-Androstene-3α, 17β-diol	0.67
	0.71
5α-Androstane-3α, 17β-diol	0.76
5α-Androstane-3β, 17β-diol	0.90
5β-Androstane-3, 17-dione	4.00
Δ^4-Androstene-3, 17-dione	6.1

[*]Chromatographed on QF-1 (3%).
[†]Cholestane 4.5 min.
Column: 6 ft by $\frac{1}{8}$ in. OD stainless steel or 6 ft by $\frac{1}{4}$ in. OD glass.

part of the column.   It would appear more reasonable to suggest that this effect might be due to the inevitable stripping of stationary phase at the top part of the column, resulting in a rapid alteration of the molecule as described before.

DECOMPOSITION OF ALLYLIC ALCOHOLS

A second group of compounds also appears to undergo decomposition irrespective of the safeguards outlined in the previous section. It has been reported that Δ^4-3-hydroxy steroids undergo elimination reaction with the formation of a diene system during gas chromatography [9].   Reduction of Δ^4-3-ketones with sodium borohydride in selective solvents generally produces a mixture of the unsaturated 3a- and 3β-compounds, the latter being produced in the largest concentration.   Chromatography of these allylic alcohols produces two peaks with retention times shorter than anticipated (Table 8-3).   Separation of the epimers on thin-layer chromatography, followed by gas chromatography of either, shows the appearance of two peaks with retention times identical for each isomer (Figure 8-5). The elimination products are obtained whether a metal injection block or an all-glass on-column injection system is used.   Nevertheless, changes in the relative peak heights, that is, the concentrations of the two isomeric dienes, are found on increasing the temperature of the injection block.

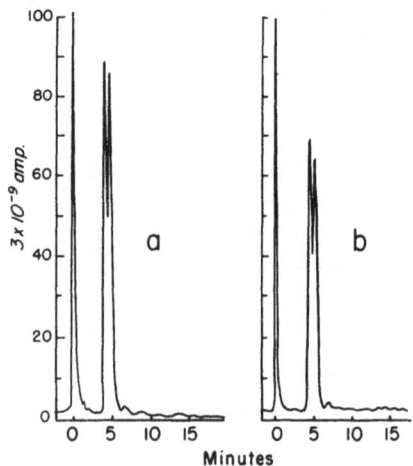

Figure 8-5. Gas–liquid chromatogram of pure (a) 3a-hydroxy-androst-4-en-17-one and (b) 3β-hydroxy-androst-4-en-17-one.   The isomers were previosly separated by TLC.   Column: 6 ft (all glass), 3.8% SE-30 on 80-100 mesh Diatoport S under 42 psi He. Temperature: column 230°C; vaporizer 280°C.

Acid treatment as reported by Stavely and Bergmann [10] produced a product with a t_r identical to the second of the pair formed on thermal elimination. This product, identified as the $\Delta^{3,5}$-diene, is the result of an acid-catalyzed rearrangement from the expected $\Delta^{2,4}$-diene. It would appear, therefore, that the substance eluted first ($r = 0.67$) is the $\Delta^{2,4}$-androstadien-17-one and the second ($r = 0.71$), the more stable $\Delta^{3,5}$-androstadien-17-one. Isolation of the $\Delta^{2,4}$-diene, followed by GLC, yields a single peak with no change in retention time, indicating that this substance is thermally stable and can be chromatographed in the vapor phase without rearrangement to the $\Delta^{3,5}$-diene.

REFERENCES

1. H. Gottfried, Steroids 5:385 (1965).
2. H. F. Martin and H. H. Wotiz, Biochim. Biophys. Acta 60:25 (1962).
3. W. J. A. VandenHeuvel, C. C. Sweeley, and E. C. Horning, J. Am. Chem. Soc. 82:3481 (1960).
4. B. Kliman, discussion of paper by E. C. Horning, Recent Progr. Hormone Res. 19:99 (1963).
5. M. B. Whittier, L. Mikkelsen, and S. Spencer, Southwestern Regional ACS Meeting, December 7, 1962.
6. J. E. Arnold and H. M. Fales, J. Gas Chromatog. 3:131 (1965).
7. W. J. A. VandenHeuvel and E. C. Horning, Biochem. Biophys. Res. Commun. 3:356 (1960).
8. C. J. W. Brooks, Anal. Chem. 37:636 (1965).
9. B. Knights, private communication.
10. H. E. Stavely and W. Bergmann, J. Org. Chem. 1:575 (1936).

Chapter 9

Irreversible Sorption

The point raised by Whittier [1] about the sorption of steroids on exposed surfaces is an extremely important one. It has been our day-to-day experience that when working quantitatively with relatively small amounts of steroid, either very small peaks or, occasionally, no peaks at all are seen following the first one or two injections. This phenomenon is caused by a form of "irreversible sorption" and requires the analyst to saturate the column each day with a few micrograms of the substances to be analyzed. Saturation can be attained by the injection of high concentrations of standards or, for the rarer compounds, through the injection of some well-preserved old extracts. When relatively high concentrations of steroid are being used, the sorption effect is frequently unnoticed. It is thought that this apparent loss of substance is caused by an interaction with polar groups at the active sites of exposed solid support, although increasing the stationary phase—thereby covering more of these sites—does not necessarily decrease the sorption. For practical purposes, of course, it is not important to know the specific area of interaction, but it is important to know that the problem can be overcome by saturation or "priming" of the column with each steroid to be analyzed.

Although the term "irreversible sorption" is loosely applied here, for practical purposes it does appear to be just that. It can be demonstrated, however, that nearly complete desorption occurs within a few hours (from 2–3 hr for the relatively heat-labile aldosterone diacetate to perhaps 8–12 hr for the estrogen acetates). It is likely that the sorbed steroid is slowly pyrolyzed, followed by elution of the fragments, requiring the renewed saturation of the binding sites. Unfortunately, silanizing has not eliminated this problem, which represents a considerable hazard in the analysis of radioactive substances especially when present in very small amounts. Loss of radioactive substances by sorption while passing through the column has not been adequately investigated. It should be pointed out that there exist considerable differences in this sorption factor depending on the nature of the compound chromatographed and the stationary

TABLE 9-1. Recovery of Estrone Acetate from an
Unprimed Column*

Injection	Amount (mg)	Recovered (mg)	Sorbed (%)	Steroid sorbed/ g packing (μg)
1	0.5	0.335	33	2.85
2	0.5	0.415	17	1.46
3	0.5	0.465	7	0.60

*Column: 3% SE-30 on Chromsorb 30-60 mesh, 3 ft by 1/2 in., total packing 58 g.
The column was cured at 250°C for 14 days prior to use and had not been used
for estrogen separation prior to this experiment.

phase. The more volatile derivatives such as the trifluoroacetates,
heptafluorobutyrates, and trimethylsilyl ethers exhibit little if any
sorption. On the other hand, free steroids, particularly if they are
polyhydroxy compounds, and to a lesser degree their acetylated derivatives, tend to show much greater sorption. In general, it has been
difficult to attain measurable peaks from steroid acetates below 0.01 –
0.02 μg even after exhaustive column saturation and despite greater
theoretical detector sensitivity. On the other hand, it has been possible, by using electron capture detectors, to chromatograph and obtain
adequate peaks for several steroid heptafluorobutyrates to concentrations of 10^{-11}g even without prior saturation of the column.

Evidence for the sorption of acetylated steroids is presented in
Table 9-1, which shows the recovery of estrone acetate following

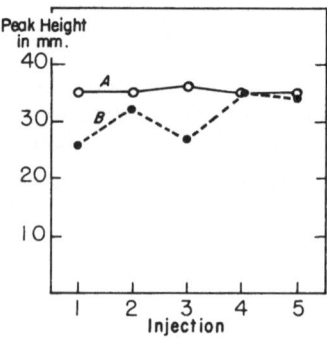

Figure 9-1. Reproducibility of injection with and without priming of the column. Curve A represents the detection response
of five consecutive injections of 0.02 μg estrone acetate after
priming the column with 12 μg estrone acetate. Curve B shows
the detection response of the same sample size without priming
column.

injection on a relatively new preparative column, coated with 3% SE-30. Collection efficiency under these circumstances was shown to be somewhat better than 90%. It is readily seen that a considerable amount of estrogen is lost on such a column following the first injection and still further loss occurs following the second injection. Figure 9-1 presents similar evidence following chromatography of estrone acetate on a well-conditioned 6-ft by $\frac{1}{8}$-in. SE-30 column. Injections were made on two successive mornings and the peak heights were plotted for the first three injections each morning. On the first morning the column was not primed, while on the second day $12 \mu g$ of estrone acetate was passed through the column in three successive injections. This was followed by three more injections of pure acetone in order to make certain that no residual material was left in the injector.

REFERENCES
1. M. B. Whittier, Med. Electronics News 4:39 (1964).

Chapter 10

Purity of Standards and Solvents

STANDARDS

Owing to the greater sensitivity of the ionization detectors, when compared to general staining techniques in paper chromatography and other procedures, one can demonstrate that virtually none of the steroids available as commercial standards produce single peaks. In most instances this is simply a reflection of the presence of a small amount of impurity, rather than decomposition during chromatography. This can be ascertained readily by trapping the main peak and re-injecting into the gas chromatograph. In a number of instances we have examined commercially available steroids in this manner and have found that only a single peak is obtained following this procedure. Almost none of the commercial steroids can be considered gas chromatographically pure. In some instances it has been quite easy to detect substances with retention times identical to those expected as by-products of the chemical reaction necessary to prepare the standard, or as biological companions where the steroid was obtained through extraction from natural sources. Indeed, the reader must be very cautious in the interpretation of the notation "chromatographically pure." This, to our knowledge, never refers to "gas chromatographically pure." In many instances, however, it is fortuitous that the amount of impurity is sufficiently small as to be of little or no consequence, provided the material is used only as a qualitative comparison standard.

SOLVENTS

Of much greater importance to the analyst is the large number of impurities found in most commercial solvents. In general, reagent grade solvents, when used for direct injection, present only one fast solvent peak. However, in the process of extraction or solvent partitioning a fairly large amount (100–500 ml) of a solvent is frequently utilized. This material is subsequently evaporated to dryness or reduced to a very small volume, and it has been our experience that numerous GLC peaks may be obtained from even the "purest" solvents

TABLE 10-1. Gas Chromatography of Distillation Residues from
Common Solvents*

Solvent	Volume evaporated (ml)	t_r (min)	r†	Peak height (mm)
Acetone	250	4.1	0.63	o.s. ‡
		5.1	0.78	o.s. ‡
(acetic anhydride and		6.4	0.99	150
pyridine treated)		8.4	1.35	85
		11.2	1.73	55
		14.4	2.23	35
		19.0	2.94	13
		24.5	3.80	6
Acetone	500	6.35	0.99	11
(singly distilled		8.50	1.32	70
spectroanalyzed)				
Benzene	100	4.4	0.64	22
(distilled)		5.3	0.82	14
		7.1	1.09	14
		9.3	1.43	100
		12.2	1.89	6
		15.6	2.48	4
Benzene	100	6.3	0.98	12
(acetic anhydride and				
pyridine treated)				
Ethyl ether	500	5.2	0.81	o.s.‡
		6.2	0.96	100
		9.1	1.41	70
		11.3	1.75	37
		15.4	2.39	16
Toluene	400	3.55	0.55	28
		5.03	0.78	15
		6.35	0.99	5
		8.55	1.32	4
		11.25	1.74	18
Petroleum ether	200	4.4	0.68	o.s.‡
		5.2	0.80	o.s.‡
		6.4	0.99	30
		8.1	1.25	88
Methylene chloride	200	No peaks		

*Column: 6 ft by $^1\!/_8$ in., 3% SE-30 on 80-100 mesh Diatoport S. Total material injected—
$^1\!/_{50}$ of final solution. Attenuation = 100×.
†Relative to estrone acetate, t_r = 6.45 min, $t_{r_{E_2A}}$ = 9.15, r_{E_2A} = 1.42.
‡Off scale.

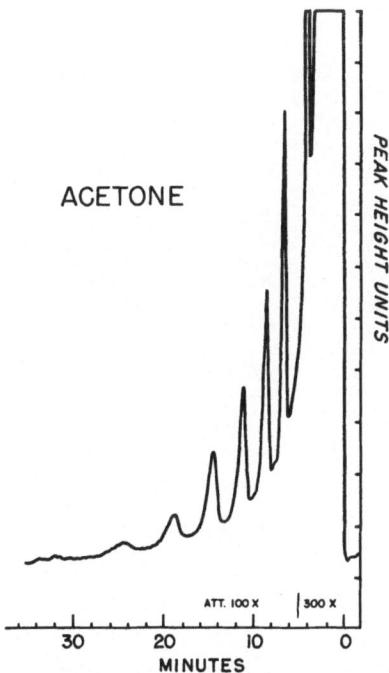

Figure 10-1. Chromatogram obtained from an injection of concentrated residue representing 10 ml of singly distilled "spectroanalyzed" acetone on 6 ft × ⅛ in. stainless steel column packed with 3% SE-30 on Diatoport S (80-100 mesh). Column conditions: 20 psi, N_2 temperature 250°C.

(i.e., spectrograde).   Figure 10-1 shows a chromatogram obtained following evaporation of 200 ml of acetone (reagent grade), dissolution of the residue in 50 μl of acetone, and injection of 2 μl of this solvent into the gas chromatograph. As can be seen, nearly all of the estrogen metabolites would probably be obscured if present in extremely low concentrations. These solvent impurities would not be noticeable in the analysis of hormones present in anything but very small amounts.

Table 10-1 shows the retention times relative to estrone acetate and the relative peak height at the various attenuations for a number of solvents before and after distillation and some before and after treatment with acetic anhydride and pyridine. In all instances, at least two, and frequently many more, peaks can be discerned. To date, we have found only one solvent which produces no extraneous peaks following evaporation of commercial grade material and injection into the gas chromatograph. This solvent is dichloromethane.

Chapter 11

Choice of Derivative

The comments made earlier about the relative sorption of derivatives suggest at first glance that only the three volatile derivatives mentioned should be used in gas chromatography. Such a viewpoint, however, would neglect the problems of separation as well as the possible difficulties attendant to the formation of the derivatives. The specific selection of the derivatives for each assay is discussed with the description of that particular procedure. However, certain general comments may be worthwhile at this time. Although the use of unreacted steroids would, of course, eliminate one time-consuming step, it has several serious drawbacks. In the first place, functional groups tend to interact with the column to the point where certain compounds, such as estriol, are so strongly sorbed that on occasion amounts of less than 1 μg cannot be detected. Moreover, adequate separations are frequently not attained with free steroids, and, as pointed out above, some compounds cannot be chromatographed except as derivatives. Despite the irreversible sorption of the acetates, we have found them to be quite useful. For example, the major estrogen acetates can be separated because of molecular-weight differences. Acetates are formed relatively quickly in quantitative yield and they are very stable in solution or dried extract over long periods of time. In some instances, such as the analysis of pregnanediol, it has been shown that adequate separation from extraneous biological material cannot be achieved by other means.

Trimethylsilyl ethers also are very useful derivatives. Certain separations, such as that of the 17-ketosteroids, cannot be achieved with any other derivative. On the other hand, the separation of the three classical estrogens as these derivatives on SE-30, XE-60, and QF-1 columns is much less secure. Evidence has been presented that the trimethylsilyl ethers are formed quantitatively and quickly, and we have found this to be true when working with pure steroids. One major drawback of these latter derivatives is the fact that they tend to hydrolyze spontaneously, catalyzed by small amounts of water or acid.

This may become troublesome when working with difficult-to-dry biological extracts.

The trifluoroacetates again are formed quantitatively in a relatively short time, and a remarkable number of excellent separations may be achieved through their use [1, 2]. On the other hand, it has been our experience that the derivatives formed are generally unstable and definite signs of decomposition are noted within 48 hr or less.

Heptafluorobutyrates have not been advocated as general derivatives, but were proposed by Clark and Wotiz [3] primarily for their excellent electron capturing ability. The limited experience we have had with these substances indicates that they allow resolution similar to that of the trifluoroacetates of many steroids, but that they are somewhat more difficult to form and maintain.

REFERENCES

1. W. J. A. VandenHeuvel, J. Sjovall, and E. C. Horning, Biochim. Biophys. Acta 48:596 (1961).
2. I. S. Hartman and H. H. Wotiz, Biochim. Biophys. Acta 90:334 (1964).
3. S. J. Clark and H. H. Wotiz, Steroids 2:535 (1963).

Chapter 12

Preliminary Purification

The discussion and presentation of gas chromatographic procedures for the analysis of steroid hormones for practical purposes must be divided into two parts.

The first deals with the rapid determination of those hormones present in relatively large concentrations (milligram amounts per twenty-four hours) which require little if any preliminary cleanup of the crude urine extract, allowing a high-efficiency gas chromatographic column to do all the necessary separating of the major components. The advantage of GLC for these procedures is obvious, since it allows the rapid and routine determination of certain substances with a minimum of effort even on the very day a specimen is obtained.

The second part treats the determination of steroid metabolites present in urine or plasma in very small concentrations—a few micrograms per twenty-four hours or fractions of a microgram in plasma. Despite the high resolution of the gas chromatographic column, the often disproportionately large peaks of interfering substances completely obscure the minor peaks representative of the hormones or hormone metabolites, making it impossible to measure them quantitatively and throwing severe doubt on the specificity of the assay. For this type of determination it becomes necessary to take advantage of a number of other preliminary purification procedures such as liquid–liquid partition, thin-layer chromatography, paper chromatography, column chromatography, and so on.

Nevertheless, a number of steroid hormones even at a low level of excretion (pregnanediol, 17-ketosteroids) can still be measured in crude extracts, basing separation only on an efficient gas chromatographic column ($\bar{H} = 0.8$ mm or less). Indeed, in one or two instances, the remarkable sensitivity of such a procedure is surprising. In our laboratory, it has appeared quite feasible to determine 17-ketosteroids and pregnanediol in a matter of 2 to 3 hr with a sensitivity not previously achieved with any other procedure involving considerably greater preliminary purification. Urinary estriol also can be measured quite rapidly in cases of pregnancy or certain types of tumors. On the other hand, the determination of the classical estrogens as well as some other estrogen metabolites in the normal female even at peak excretion periods of the menstrual cycle needs an extensive preliminary cleanup.

Chapter 13

Relationship of Chromatographic Behavior and Steroid Structure

(by I. S. Hartman)

The structural analysis of steroids by gas chromatography has been developed via two general approaches:

1. The study of the characteristic changes in retention times (Δr)* subsequent to changes in functional groups.
2. The study of the relationship between functional groups and the nature of the stationary phase (T values and steroid number).

MEASUREMENT OF RETENTION TIME AND CALCULATION OF r VALUES

Martin [1] has developed a theory of partition coefficients based on the assumption that the chemical potential of a substance is an additive function of the constituent parts of its molecule. If μ is the chemical potential of the substance A and if $\Delta\mu_A$ is equal to the free energy required to transport one mole of A from one phase to another, then as a first approximation $\Delta\mu_A$ may be regarded as being equal to the sum of the potential differences $(\Delta\mu_a, \Delta\mu_b)$ of the various groups which make up molecule A:

$$\Delta\mu_A = \Delta\mu_a + \Delta\mu_b + \Delta\mu_c$$

Martin's concept of the additive contribution to chromatographic mobility of the components of a molecule has been recently applied to the behavior of steroid molecules during gas–liquid chromatography.

To a first approximation, the free energy required to transport a given group (a) from one solvent to another is independent of the rest of the molecule. The addition of group (a) to a molecule will change the partition coefficient of a substance, and hence its mobility

---

*r is the relative retention of a compound and Δr represents the change in r upon a particular change in the molecule; hence, Δr is the r value of the substituent group.

99

and retention time, by a definite amount in a particular solvent system, or stationary phase. This change will be dependent on the nature
of the group (a) and independent of the structural features of the rest
of the molecule [1]. Introduction of more than one (a) group into a
molecule will cause a change in the partition coefficients directly
proportional to the number of groups of (a). England and Cohn [2]
have demonstrated a regular change in the partition coefficients of a
homologous series of amino acids with the introduction of each carbon atom into the chain. This finding is indicative of a corresponding
change in the R_f values of the members of this homologous series.

In pointing out the limitations of the applicability of his theory,
Martin had suggested that stereochemical factors might be a major
cause of deviation from theory in the relationship of structure to
partition coefficients. Owing to steric factors, the solvent molecules
may be unable to approach the solute molecules and the energy of
association between the molecules of the solute and the solvent will be
lower than expected from a consideration of the sum of various groups
of the solute molecules. Bate-Smith and Westall [3] have reported
excellent agreement with the theory in their study of a wide range of
polyphenolic compounds with substituents where little or no steric
hindrance was involved.

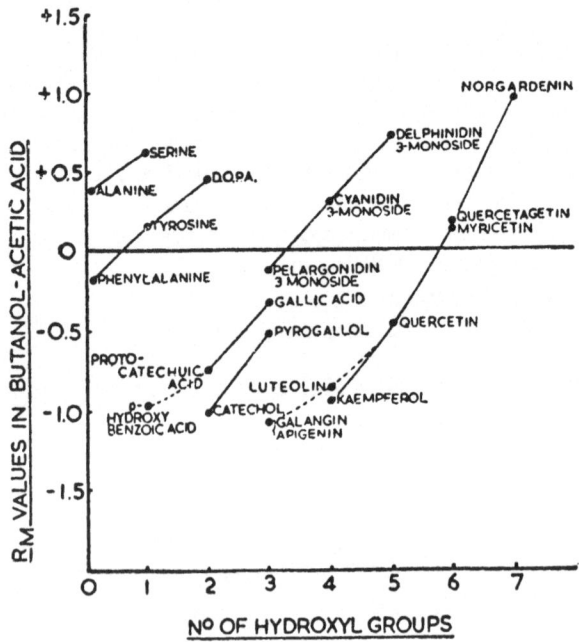

Figure 13-1. R_M [$= \log (1/R_f - 1)$] values in acid butanol [E. C. Bate-Smith and R. G. Westall,
Biochim. Biophys. Acta 4:427 (1950)].

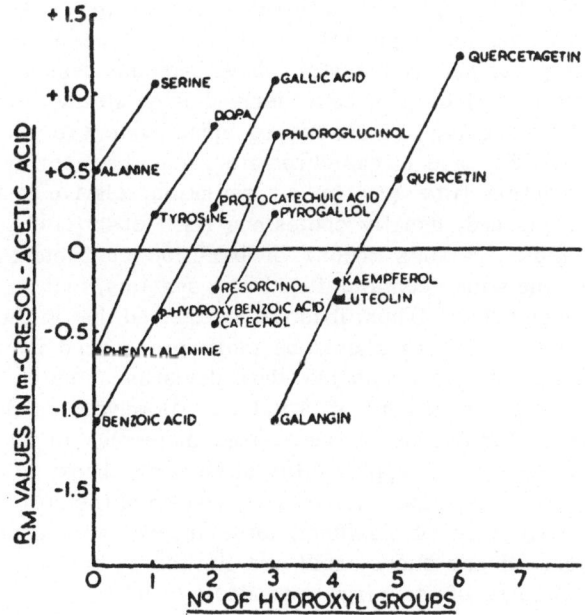

Figure 13-2. R_M [$= \log (1/R_f - 1)$] values in acid butanol [E. C. Bate-Smith and R. G. Westall, Biochim. Biophys. Acta 4:427 (1950)].

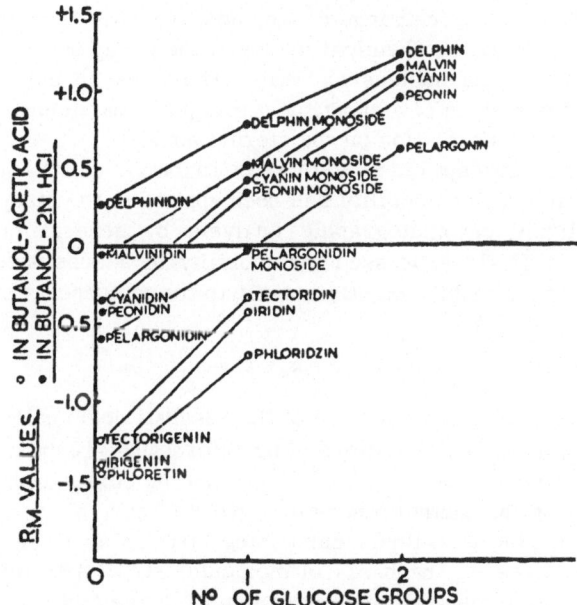

Figure 13-3. R_M [$= \log (1/R_f - 1)$] values in acid butanol [E. C. Bate-Smith and R. G. Westall, Biochim. Biophys. Acta 4:427 (1950)].

The applicability of Martin's theory to steroid molecules has been reviewed and the developments of R (or r) values with particular reference to paper partition systems have been discussed by Bush [4]. R is a function which has been derived in an attempt to relate the mobility of a molecule in a chromatographic system to its molecular structure. In the case of gas chromatography the mobility is measured as retention time (t_r) or as retention relative to a standard reference compound, usually cholestane (r). Bush [4] observes that deviations from Martin's theory of partition coefficients begin to appear when the solute contains fused ring systems, such as are found in steroid molecules. Thus, it is not unexpected that in this class of compounds steric factors should be more pronounced and that more of the substituent groups should show deviation from theory in the prediction and measurement of R values. However, as Bush points out, the steric limitations in the steroid molecules might be turned into an advantage in the applicability of Martin's theory to this group of compounds. The rigidly joined ring system of the steroid nucleus has little possibility of conformational inversions. Consequently, conformations of any one steroid are limited in number. Furthermore, substituents of the large extended steroid molecule are sufficiently far apart so as to exert little or no influence on one another. This is an ideal requirement for the applicability of the theory. Bush has made the practical suggestion that in order to make the theory applicable over a wide range of steroids, the word "group" be defined to mean not only the chemical nature of the group but its position and orientation on the nucleus as well. Thus, Δr should represent the retention time value of a substituent group at a particular position and a particular conformation on the steroid nucleus.

Martin's concept of the additive contribution of substituent groups to chromatographic mobility has been applied to the study of steroids in gas–liquid chromatographic analysis by several investigators.

Clayton [5–7] expressed the mobility of a substituted molecule in terms of the mobilities attributable to the components of that molecule:

$$r = r_n + k_a + k_b + \ldots$$

where r_n is the retention time of the unsubstituted nucleus and k_a and k_b are group retention factors of noninteracting groups at positions a and b on the nucleus. Clayton [5] studied the correlation between the structure of the methyl ethers of a group of neutral sterols and their behavior during gas–liquid chromatography in an attempt to predict retention times on the basis of molecular structure. He interpreted his data in terms of the operation of three factors: (a) polarity, (b) molecular weight, (c) conformational restriction; and derived retention constants for the (C=C) functional groups (Table 13-1) [5].

TABLE 13-1. Retention Ratios for Unsaturated Structures of Different Molecular Weight*

Retention ratio†	Cholestanol	Ergostanol	Stigmastanol	Lanostanol
Δ^7/stanol	1.30	1.26	‡	‡
$\Delta^{5,7}$/stanol	1.52	1.52	‡	‡
$\Delta^{5,7}/\Delta^7$	1.17	1.20	‡	‡
Δ^5/stanol	1.08	1.07	1.09	‡
$\Delta^{5,22}/\Delta^5$	0.98	0.87	0.88	‡
$\Delta^{5,7,22}/\Delta^{5,7}$	‡	0.89	‡	‡
Δ^{22}/stanol	‡	‡	0.90	‡
$\Delta^{8,24}/\Delta^8$	‡	‡	‡	1.33
$\Delta^{5,24}/\Delta^5$	1.38	‡	‡	‡
$\Delta^{5,24(28)}/\Delta^5$	‡	1.18	1.19	‡

*Reproduced by permission from R. B. Clayton, Nature 190:1071-1072 (1961).

†Retention ratio = $\dfrac{\text{relative retention of compound with substituent}}{\text{relative retention of analogue lacking substituent}}$

‡Compounds necessary for comparison not available.

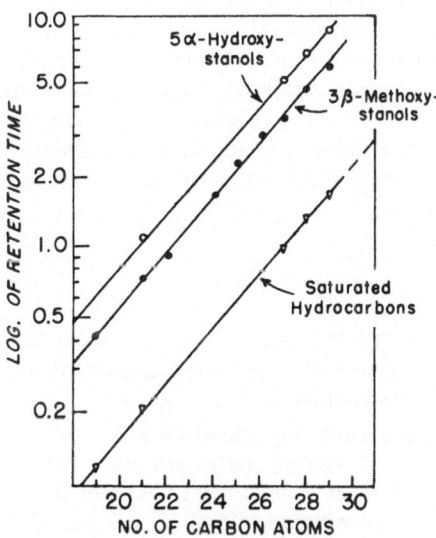

Figure 13-4. Relative retention times of some steriod hydrocarbons and monosubstituted derivatives [R. B. Clayton, Biochemistry 1:357-366 (1962)].

TABLE 13-2. Log r Contributions for C-10 Methyl Group*

Compound	r	Log r (C−10 methyl)
17β-Hydroxyandrost-4-en-3-one	4.24	0.08
17β-Hydroxy-17α-methylandrost-4-en-3-one	4.30	0.06
17β-Hydroxy-17α-ethinylandrost-4-en-3-one	3.92	0.07
Pregn-4-ene-3,20-dione	8.62	0.07
20α-Hydroxypregn-4-en-3-one	6.63	0.07
20β-Hydroxypregn-4-en-3-one	5.91	0.07

*Reproduced by permission from B. A. Knights and G. H. Thomas, Anal. Chem. 34:1046-1048 (1962).

Clayton's later observation [6] suggests that the relationship expressed in the above equation is a general one such that the introduction of a substituent into a particular position on the steroid nucleus contributes to the retention time of the whole molecule by a factor which is a constant both for the type of grouping, and (as defined by Bush [4]) for its particular site and conformation, but is independent of the molecular weight of the total structure and the influence of other functional groups with which there is no interaction. It was also expressed by Clayton [6] that the actual values of the group retention factors would differ greatly depending on the chemical nature of the liquid phase. Clayton has observed [7], in agreement with his early observation, a straight line relationship between the logarithm of the retention time of a series of 3β-methoxy-5α compounds and the number of carbon atoms in the molecule (Figure 13-4). The straight line relationship is not exact, however, which is not unexpected since the compounds are not members of a true homologous series.

Knights and Thomas [8] have used Clayton's equation* in its logarithmic form:

$$\log r = \log r_n + \log k_a + \log k_b + \ldots$$

They investigated a large number of steroids and demonstrated that in the absence of vicinal effects, log r for a particular group was constant and usually independent of other groups in the molecule. Data in Table 13-2 demonstrate this constancy for the contribution of the C-10 methyl group measured on the liquid phase QF-1.

Knights and Thomas have observed considerable variations in the log r values for an 11β-hydroxyl and an 11-ketone (Table 13-3), which they find indicative of the influence of the neighboring groups on the

*See p. 102.

TABLE 13-3.   Log r Values for Ketone, Hydroxyl, and Methyl
Substituents*

Substituent	Compound	r	Log r (substituent)
11-Ketone	5β-Androstane-3,11,17-trione	7.41	0.24
	5α-Pregnan-11-one	0.79	0.48
	5α-Pregnane-3,11,20-trione	12.42	0.32
	5β-Pregnane-3,11,20-trione	11.10	0.32
	Pregn-4-ene-3,11,20-trione	17.40	0.30
11β-Hydroxyl	3α, 11β-Dihydroxy-5β-androstan-17-one	4.19	0.27
	5α-Pregnan-11β-ol	0.71	0.43
11α-Hydroxyl	11α-Hydroxypregn-4-ene-3,20-dione	17.65	0.31
12α-Hydroxyl	3α,12α-Dihydroxy-5β-pregnan-20-one	4.31	0.20
17α-Hydroxyl	3β, 17α-Dihydroxypregn-5-en-20-one	4.19	0.19
	17α-Hydroxypregn-4-ene-3,20-dione	13.00	0.18
16-Methyl	16α-Methylpregn-4-ene-3,20-dione	8.44	−0.01
	16β-Methylpregn-4-ene-3,20-dione	8.52	−0.01

*Reproduced by permission from B. A. Knights and G. H. Thomas, Anal. Chem. 34:1046–
1048 (1962), Table III.

TABLE 13-4.   Log r Values for
Compounds   Chromatographed   on
QF-1 at 250°*

Substituent	Log r
3-OH (equatorial, 5α)	0.47
3-OH (equatorial, 5β)	0.49
3-OH (axial, 5α)	0.43
3-OH (axial, 5β)	0.44
17β-OH	0.44
20α-OH	0.45
20β-OH	0.40
3-Ketone	0.75
17-Ketone	0.65
20-Ketone	0.57

*Reproduced by permission from B. A. Knights
and G. H. Thomas, Anal. Chem. 34:1046-1048
(1962).

TABLE 13-5.   Relative Retention Data for Free and Derivatized Androstanes*

Steroids	SE-30 3%			XE-60 3%			NGSeb 3%			Hi-Eff 8B 3%		
	Free	Acetate	TMSi	Free	Acetate	TMSi	Free	Acetate	TMSi	Free	Acetate	TMSi
5α-Androstane	0.21	—	—	0.27	—	—	0.09	—	—	0.13	—	—
5β-Androstane	0.20	—	—	0.25	—	—	0.08	—	—	0.12	—	—
5α-Androstan-17β-ol	0.32	0.39	0.33	0.60	0.71	0.38	0.47	0.44	0.20	0.71	0.58	0.22
5β-Androstan-17β-ol	0.41	0.52	0.49	0.89	1.04	0.52	0.84	0.80	—	1.48	1.27	—
5α-Androstan-17-one	0.32	—	—	0.65	—	—	0.39	—	—	0.61	—	—
5β-Androstan-17-one	0.42	—	—	0.95	—	—	0.71	—	—	1.15	—	—
5α-Androstan-3β-ol	0.31	0.39	0.34	0.65	0.70	0.39	0.47	0.47	0.21	0.75	0.71	0.24
5β-Androstan-3β-ol	0.29	0.35	0.31	0.53	0.59	0.32	0.38	0.37	—	0.61	0.57	—
5α-Androstan-3-one	0.33	—	—	0.77	—	—	0.49	—	—	0.73	—	—
5β-Androstan-3-one	0.32	—	—	0.68	—	—	0.42	—	—	0.67	—	—
5α-Androstane-3α,17β-diol	0.54	0.89	0.58	1.67	2.05	0.49	2.27	1.69	0.31	3.94	2.48	0.28
5β-Androstane-3α,17β-diol	0.51	0.87	0.61	1.68	1.95	0.50	2.08	1.89	0.35	3.85	2.82	0.37
5α-Androstane-3β,17β-diol	0.59	0.98	0.69	1.89	2.21	0.66	2.44	2.20	0.44	4.19	3.18	0.42
5β-Androstane-3β,17β-diol	0.49	0.86	0.56	1.56	1.90	0.51	1.93	1.66	0.28	3.50	2.42	0.28
5α-Androstan-3α-ol,17-one	0.51	0.58	0.47	2.00	2.14	0.87	1.83	1.61	0.58	3.73	2.62	0.74
5β-Androstan-3α-ol,17-one	0.47	0.56	0.46	1.90	2.15	0.94	1.80	1.76	0.72	3.58	2.86	1.03
5α-Androstan-3β-ol,17-one	0.55	0.66	0.55	2.10	2.44	1.16	2.10	2.00	0.91	3.97	3.44	1.20
5β-Androstan-3β-ol,17-one	—	—	—	1.98	2.98	0.86	1.88	—	—	—	—	—
5α-Androstan-3-on,17β-ol	0.59	0.76	0.64	2.58	2.81	1.32	2.61	2.38	1.07	4.61	3.88	1.27
5β-Androstan-3-on,17β-ol	0.54	0.69	0.59	2.44	2.53	1.18	2.20	1.98	0.91	4.18	3.44	1.13
5α-Androstan-3-on,17σ-ol	0.56	0.75	0.60	2.03	2.32	1.16	2.22	2.12	0.94	4.36	3.31	1.09
5β-Androstan-3-on,17α-ol	0.53	0.61	0.47	2.14	2.14	0.91	2.05	1.76	0.62	3.97	2.91	0.81
5α-Androstane-3,17-dione	0.58	—	—	2.64	—	—	2.27	—	—	4.00	—	—
5β-Androstane-3,17-dione	0.50	—	—	2.48	—	—	1.98	—	—	3.64	—	—
Androst-5-en-3β-ol,17-one	0.61	0.64	0.55	2.01	2.38	1.10	2.09	1.99	0.90	4.00	3.38	1.16

*Reproduced by permission from I. S. Hartman and H. H. Wotiz, Biochim. Biophys. Acta 90:334–348 (1964), Table I.

TABLE 13-6. Comparison of Retention Data of Isomeric Pairs:
Substituent Effect [*]

Functional group	SE-30 3%		XE-60 3%		NGSeb 3%		Hi-Eff 8B 3%	
	5α	5β	5α	5β	5α	5β	5α	5β
No substitution	0.21	0.20	0.27	0.25	0.09	0.08	0.13	0.12
17β-ol	0.32	0.41	0.60	0.89	0.47	0.84	0.71	1.48
3β-ol	0.31	0.29	0.65	0.53	0.47	0.38	0.75	0.61
3α,17β-diol	0.54	0.51	1.67	1.68	2.27	2.08	3.94	3.85
3β,17β-diol	0.59	0.49	1.89	1.56	2.44	1.93	4.19	3.18
17-one	0.32	0.42	0.65	0.95	0.39	0.71	0.61	1.15
3-one	0.33	0.32	0.77	0.68	0.49	0.42	0.73	0.67
3,17-dione	0.58	0.50	2.64	2.48	2.27	1.98	4.00	3.64
3α-ol,17-one	0.51	0.47	2.00	1.90	1.83	1.80	3.73	3.58
3β-ol,17-one	0.55	—	2.10	1.98	2.10	1.88	3.97	—
3-one,17β-ol	0.59	0.54	2.58	2.44	2.61	2.20	4.61	4.18
3-one,17α-ol	0.56	0.53	2.09	2.14	2.22	2.05	4.36	3.97

[*]Reproduced by permission from I.S. Hartman and H.H. Wotiz, Biochim. Biophys. Acta
90:334-348 (1964), Table II.

log r contribution of the substituent [8]. Table 13-4 summarizes the
log r values obtained by these two investigators for the functional
groups commonly found in steroids.

Hartman and Wotiz [9, 10] have studied the gas—liquid chromatographic behavior of a series of isomeric C-19 steroid pairs and
have calculated log r values for the hydroxyl, ketone, acetyl, and
trimethylsilyl groups, and have drawn correlations between gas—
liquid chromatographic behavior of the steroids and the chemical
nature of the liquid phase (Tables 13-5 and 13-6).

Certain deductions concerning the polarities of the four stationary
phases studied by Hartman and Wotiz may be drawn from the relative
retention of the steroids and their three derivatives (Table 13-5).
The general pattern of the retention time values is: free steroid[*] >
acetate > trimethylsilyl for the two polyester columns NGSeb and
Hi-Eff 8B; acetate > free steroid > trimethylsilyl for the XE-60
polymer; and acetate > trimethylsilyl ≃ free steroid for the SE-30
elastomer. The polarities of the phase may thus be approximated
as Hi-Eff 8B ≃ NGSeb > XE-60 > SE-30.

[*]Meaning free from a substituent on the hydroxyl group.

In Table 13-6 it is seen that on all columns studied the retention time of the 5 α-androstane nucleus is consistently greater than that of the 5β-isomer. Although this difference is very small, it is nevertheless constant and reproducible on all the columns studied and is reflected in the retention times of the oxygenated nuclei as well, indicating a constant contribution to retention time of the 5α-configuration over the 5β.

A comparison of the relative retention data of the C-3 oxygenated steroids in Tables 13-5 and 13-6 indicates a further correlation between chromatographic behavior and conformation. In the 5β-series of the C-19 steroids, the 3β-hydroxy isomers (axial conformation) exhibit greater mobility than the corresponding 3α-hydroxy compounds (equatorial conformation). In the 5α-series, on the other hand, this pattern is reversed and compounds with the hydroxyl group in the 3α-orientation (axial) exhibit greater mobility than their 3β-isomers (equatorial). These observations are in agreement with those of Savard [11] on the paper partition chromatography of C-19 and C-21 steroids and with the general concept defined by Barton [12, 13]. Brooks and Hanaineh [14] have suggested that the elution of the 3α-hydroxy-5α-androstan-17-one acetate before its 5β-isomer might be due to the selective retention of the equatorial acetoxy groups by the polar polyester liquid phase. The findings of Hartman and Wotiz [10] are in agreement with those of Knights and Thomas [8], who have

TABLE 13-7. Effect of Substituents at C-3 on Resolution of 5α- and 5β-Steroids*

Substituent	Compound	r	ΔLog r (5$\alpha\rightarrow$5β)
3-Ketone	5α-Cholestan-3-one	5.46	-0.04
	5α-Androstan-3-one	1.02	-0.04
	5α-Androstane-3,17-dione	4.72	-0.04
	5α-Pregnane-3,20-dione	5.90	-0.04
	5α-Pregnane-3,11,20-trione	12.42	-0.04
3-Equatorial hydroxyl	5α-Cholestan-3β-ol	3.03	-0.05
	5α-Androstan-3β-ol	0.54	-0.04
	5α-Pregnane-3β,20α-diol	2.25	-0.03
	5α-Pregnane-3β,20β-diol	2.02	-0.04
3-Equatorial acetate	3β-Acetoxy-5α-cholestane	4.36	-0.08
	3β-Acetoxy-5α-androstane	0.79	-0.07
	3β,20α-Diacetoxy-5α-pregnane	5.06	-0.08
	3β,20β-Diacetoxy-5α-pregnane	4.71	-0.08

*Reproduced by permission from B. A. Knights and G. H. Thomas, Anal. Chem. 34:1046-1048 (1962), Table VI.

also observed that 5β-compounds are always eluted ahead of their 5α-isomers and that the degree of separation of the two was influenced by the nature of the substitution of C-3 (Table 13-7).

In Table 13-5 it is seen that the acetyl and the trimethylsilyl ether derivatives follow the order of retention of the free steroid C-5 isomeric pairs ($5\alpha/5\beta$ pair) in all but a few instances. However, the retention times of the isomeric derivatives do not always retain the same ratio to each other as do the corresponding free compounds. A similar observation was made by Lipsky and Landowne [15], who reported that the observed decreases in retention times of steroid acetates on polar columns were not the same in all cases.

Brooks and Hanaineh [14] have carried out an impressive study of the gas chromatographic behavior of a large number of steroids and have calculated retention factors of various substituent groups. They demonstrated, once again, the validity of Martin's theory of partition coefficients. The regularities observed by Brooks and Hanaineh for a number of C-19, C-21, C-24, C-27, and C-28 steroids with the silicone phases SE-30 and QF-1 are shown in Tables 13-8 and 13-9. (All values are relative to cholestane.) Similar regularities in gas–liquid chromatography behavior can be found in the data of Hartman and Wotiz [10] (Table 13-10) and Clayton [7] (Table 13-11).

The separation factors of the $5\alpha/5\beta$ isomeric steroid pairs of the substituted and the unsubstituted compounds as calculated by Brooks and Hanaineh [14] are given in Table 13-12.

TABLE 13-8. Relative Retention Data of Steroids at 200° *

Steroid	Relative retention	
	SE-30	QF-1
Androstanes:		
5α-Androstane	0.078	—
5α-Androstan-3-one	0.175	0.805
5α-Androstan-17-one	0.161	0.622
5α-Androstan-3β-ol	0.163	0.427
5α-Androstan-17α-ol	0.163	0.368
5α-Androstan-17β-ol	0.166	0.388
5α-Androstane-3,17-dione	0.366	4.649
3α-Hydroxy-5α-androstan-17-one	0.336	1.957
3β-Hydroxy-5α-androstan-17-one	0.338	2.240
17β-Hydroxy-5α-androstan-3-one	0.386	2.630
5α-Androstane-3β,17β-diol	0.353	1.344
5α-Androstane-3,11,17-trione	0.475	9.036
5β-Androstane	0.071	—

TABLE 13-8. Continued

Steroid	Relative retention	
	SE-30	QF-1
5β-Androstan-3α-ol	0.147	0.388
5β-Androstane-3,17-dione	0.328	4.306
3α-Hydroxy-5β-androstan-17-one	0.306	2.051
17β-Hydroxy-5β-androstan-3-one	0.344	2.422
5β-Androstane-3α,17β-diol	0.322	1.202
5β-Androstane-3,11,17-trione	0.410	8.043
3α-Hydroxy-5β-androstane-11,17-dione	0.392	3.984
5α-Androst-2-en-17-one	0.155	0.631
Androst-4-en-3-one	0.221	1.287
Androst-4-ene-3,17-dione	0.452	7.247
17α-Hydroxyandrost-4-en-3-one	0.478	3.893
17β-Hydroxyandrost-4-en-3-one	0.483	4.144
Androst-4-ene-3,6,17-trione	0.642	—
Androst-4-ene-3,11,17-trione	0.555	12.64
11β-Hydroxyandrost-4-ene-3,17-dione	0.822	—
Androst-5-en-3β-ol	0.163	0.396
3β-Hydroxyandrost-5-en-17-one	0.332	1.990
Androst-5-en-17β-ol	0.167	0.374
Androst-5-ene-3β,17β-diol	0.353	1.242
Pregnanes:		
5α-Pregnane	0.148	0.206
5α-Pregnan-3-one	0.335	1.381
5α-Pregnan-11-one	0.201	0.536
5α-Pregnan-20-one	0.278	0.861
5α-Pregnan-3α-ol	0.304	0.647
5α-Pregnan-3β-ol	0.311	0.733
5α-Pregnan-20α-ol	0.305	0.642
5α-Pregnan-20β-ol	0.282	0.566
5α-Pregnane-3,20-dione	0.636	6.295
3β-Hydroxy-5α-pregnan-20-one	0.593	3.143
20β-Hydroxy-5α-pregnan-3-one	0.644	3.837
5α-Pregnane-3β,20α-diol	0.652	2.268
5α-Pregnane-3β,20β-diol	0.598	2.010
5α-Pregnane-3,11,20-trione	0.884	15.37
3β-Hydroxy-5α-pregnane-11,20-dione	0.856	8.396
5β-Pregnane	0.134	0.181
5β-Pregnan-20-one	0.251	0.779
5β-Pregnan-3α-ol	0.283	0.653
5β-Pregnan-3β-ol	0.279	0.574
5β-Pregnane-3,20-dione	0.568	5.754
3α-Hydroxy-5β-pregnan-20-one	0.531	2.821
3β-Hydroxy-5β-pregnan-20-one	0.523	2.486
5β-Pregnane-3α,20α-diol	0.589	2.065
5β-Pregnane-3α,20β-diol	0.544	1.806

TABLE 13-8. Continued

Steroid	Relative retention	
	SE-30	QF-1
5β-Pregnane-3,11,20-trione	0.769	13.38
5α-Pregnan-2-en-20-one	0.272	0.876
5α-Pregn-2-ene-11,20-dione	0.396	2.330
11α-Hydroxy-5α-pregn-2-en-20-one	0.484	2.166
16α-Methyl-5α-pregn-2-en-20-one	0.275	0.832
16α-Methyl-5α-pregn-2-ene-11,20-dione	0.387	2.132
Pregn-4-en-3-one	0.420	2.191
Pregn-4-ene-3,20-dione	0.791	9.936
Pregn-4-ene-3,6,20-trione	1.173	—
11α-Hydroxypregn-4-ene-3,20-dione	1.510	—
11β-Hydroxypregn-4-ene-3,20-dione	1.469	—
3β-Hydroxypregn-5-en-20-one	0.585	2.831
Cholanes:		
5α-Cholane	0.403	0.475
5α-Cholan-12-one	0.721	1.857
5α-Cholan-12α-ol	0.706	1.096
5β-Cholane	0.365	0.431
5β-Cholan-12-one	0.648	1.581
5β-Cholan-12α-ol	0.618	0.918
Cholestanes:		
5α-Cholestane	1.000	1.000
5α-Cholestan-3-one	2.291	6.953
5α-Cholestan-6-one	2.049	—
5α-Cholestan-3α-ol	2.120	3.237
5α-Cholestan-3β-ol	2.135	3.662
5α-Cholestane-3,6-dione	4.223	—
5β-Cholestane	0.903	0.897
5β-Cholestan-3-one	2.055	6.277
5β-Cholestan-3α-ol	1.918	3.240
5β-Cholestan-3β-ol	1.892	2.829
5α-Cholest-2-ene	—	1.006
Cholest-4-en-3-one	2.891	11.25
Cholest-4-ene-3,6-dione	4.414	—
Cholest-5-ene	0.999	0.967
Cholest-5-en-3-one	—	11.18
Cholest-5-en-3β-ol	2.120	3.327

*Reproduced by permission from C. J. W. Brooks and L. Hanaineh, Biochem. J. 87:151-161 (1963), Table I.
Relative retention values (cholestane = 1.00) are means of at least two determinations on each column and are cited to three decimal places to avoid rounding off errors in the calculation of retention factors. Retention time of cholestane: 30-33 min on SE-30, 5.5—6.0 min on QF-1. Asymmetric peaks are underlined.

TABLE 13-9. Changes in Relative Retention Accompanying the Introduction of the 3β-Hydroxyl Group into Steroids of the Androstane, Pregnane, Cholestane, and Ergostane Series *

Steroid	225° Relative retention			200° Relative retention		
	3β-Hydroxy steroid	Unsubstituted steroid	Factor due to 3β-hydroxyl group	3β-Hydroxy steroid	Unsubstituted steroid	Factor due to 3β-hydroxyl group
5α-Androstane	—	—	—	0.163	0.078	2.09
5α-Androstan-17-one	0.40	0.20	1.96	0.338	0.161	2.10
5α-Androstan-17β-ol	0.41	0.21	1.93	0.353	0.166	2.12
Androst-5-en-17β-ol	0.40	0.21	1.94	0.353	0.167	2.12
5α-Pregnane	—	—	—	0.311	0.148	2.10
5α-Pregnan-20-one	0.64	0.33	1.96	0.593	0.278	2.13
5α-Pregnan-20α-ol	0.70	0.35	2.00	0.652	0.305	2.14
5α-Pregnan-20β-ol	0.65	0.33	1.95	0.598	0.282	2.12
5β-Pregnane	—	—	—	0.279	0.134	2.09
5β-Pregnan-20-one	0.58	0.30	1.90	0.523	0.251	2.09
5α-Cholestane	1.97	1.00	1.97	2.135	1.000	2.14
5β-Cholestane	1.76	0.91	1.94	1.892	0.903	2.10
Cholest-5-ene	1.96	1.00	1.96	2.120	0.999	2.12
5α-Cholestan-6-one	3.76	1.93	1.95	—	—	—
5α-Ergost-7-ene	2.94	1.47	2.00	—	—	—
5α-Ergost-8(14)-ene	2.54	1.28	1.98	—	—	—
5α-Ergosta-7,22-diene	2.48	1.25	1.98	—	—	—
Mean values:			1.96			2.11
(standard deviation:			0.03)			0.02)

*Reproduced by permission from C. J. W. Brooks and L. Hanaineh, Biochem. J. 87:151–161 (1963), Table III. The column used was 1% SE–30 on Gas Chrom P. The retention data are expressed relative to cholestane (11–12 min at 225°; 32–33 min at 200°) = 1.00.

The practical aspects of the calculations of separation factors have been discussed by Hartman and Wotiz [9, 10], who have utilized the greater separation factors observed among the trimethylsilyl derivatives of the isomeric androstanes to develop a method for the separation of these steroids as the trimethylsilyl derivatives (Table 13-13).

The principal aims of the previously mentioned investigators and many others in testing Martin's theory, were to assess the regularities of the behavior of the steroids during gas chromatographic analysis; to correlate the retention data thus obtained with the structural features of the compounds; and to utilize these values in the prediction of retention values and in the identification, separation, and eventual assay of these compounds using gas–liquid chromatography. Thus it is very encouraging that a comparison of observed and calculated retention time values of both Clayton [6] and Hartman and Wotiz [10] are in close agreement with each other and with the theory that non-interacting groups contribute to the mobility of a compound by an amount independent of the rest of the molecule (Tables 13-14 and 13-15).

The retention time of a solute in the chromatographic column is the sum of its residence time in each phase of the column and is dependent on the degree of interaction of the solute and the solvent. The molecular weight and the polarity of a solute are factors affecting the retention time and may act in a synergistic or antagonistic manner in determining the retention time. A substituent such as the acetyl group, while increasing the molecular weight, decreases the polarity due to the hydroxyl group and contributes to a net increase of retention time of the steroid acetate on nonpolar columns, and to a decreased value on polar columns. In the case of the trimethylsilyl ether group, the great reduction in retention times observed on polar columns is probably due to the virtual elimination of polarity as a result of tight bonding and heavy shielding of the ether linkage.

The retention factors due to acetylation and ether formation have been studied by Knights and Thomas [8, 16], Hartman and Wotiz [10], Brooks and Hanaineh [14], and Martin and Wotiz [17]. Knights and Thomas [16] have investigated the characteristic log r values for the acetylation of 3- and 20-hydroxyl groups. These workers also obtained such values for trimethylsilation and were able to characterize C-3 equatorial and axial groups in the 5α - and 5β-series by gas chromatography of these derivatives (Table 13-16).

Retention factors due to acetylation of the 3-hydroxyl group were calculated by Brooks and Hanaineh [14] for SE-30 and QF-1 columns; their data are shown in Table 13-17.

Hartman and Wotiz [10] calculated the average change in retention

TABLE 13-10. Calculated Group Retention Factors ($\log r = \log r_n + \log k_a + \log k_b$)

Functional group	SE-30 3% 5α			SE-30 3% 5β			XE-60 3% 5α			XE-60 3% 5β		
	Free	Acetate	TMSi	Free	Acetate	TMSi	Free	Acetate	TMSi	Free	Acetate	TMSi
17β-ol												
17β-ol/androstane	0.19	0.28	0.20	0.31	0.41		0.35	0.42	0.14	0.55	0.62	0.32
3β,17β-diol/3β-ol	0.27	0.40	0.31	0.23	0.39		0.41	0.50	0.23	0.47	0.51	0.20
3-one,17β-ol/3-one	0.25	0.36	0.29	0.23	0.34		0.51	0.63	0.23	0.55	0.57	0.24
17α-ol												
3-one,17α-ol/3-one	0.23	0.36	0.26	0.22	0.28		0.42	0.48	0.18	0.50	0.05	0.13
3β-ol												
3β-ol/androstane	0.18	0.28	0.22	0.16	0.25		0.38	0.41	0.16	0.33	0.37	0.11
3β,17β-diol/17β-ol	0.26	0.40	0.32	0.08	0.22		0.45	0.49	0.25	0.24	0.26	-0.01
3β-ol,17-one/17-one	0.24	0.31	0.23	–	–		0.51	0.57	0.25	0.32	0.50	-0.04
3α-ol												
3α,17β-diol/17β-ol	0.22	0.35	0.25	0.10	0.23		0.44	0.46	0.12	0.28	0.27	-0.02
3α-ol,17-one/17-one	0.21	0.26	0.17	0.05	0.13		0.49	0.47	0.07	0.30	0.35	0.00
17-one												
17-one/androstane	0.19	–	–	0.32	0.32		0.38	–	–	0.58	–	–
3β-ol,17-one/3β-ol	0.25	0.22	0.20	–	–		0.51	0.54	0.47	0.57	0.70	0.43
3,17-dione/3-one	0.24	–	–	0.20	–		0.53	–	–	0.56	–	–
3-one												
3-one/androstane	0.20	–	–	0.20	–		0.45	–	–	0.43	–	–
3-one,17β-ol/17β-ol	0.26	0.29	0.29	0.12	0.13		0.62	0.60	0.55	0.44	0.38	0.36
3,17-dione/17-one	0.26	–	–	0.08	–		0.61	–	–	0.42	–	–

Functional group	NGSeb 3%						Hi-Eff 8B 3%					
	5α			5β			5α			5β		
	Free	Acetate	TMSi	Free	Acetate	TMSi	Free	Acetate	TMSi	Free	Acetate	TMSi
17β-ol												
17β-ol/androstane	0.72	0.69	0.35	1.02	1.00	—	0.74	0.66	0.23	1.09	1.02	—
3β,17β-diol/3β-ol	0.71	0.67	0.32	0.71	0.65	—	0.75	0.65	0.26	0.76	0.63	—
3-one,17β-ol/3-one	0.73	0.69	0.34	0.72	0.67	0.34	0.80	0.73	0.24	0.79	0.71	0.23
17α-ol												
3-one, 17α-ol/3-one	0.66	0.64	0.28	0.69	0.62	0.17	0.78	0.66	0.18	0.77	0.64	0.08
3β-ol												
3β-ol/androstane	0.72	0.72	0.37	0.68	0.66	—	0.77	0.74	0.27	0.70	0.67	—
3β,17β-diol/17β-ol	0.71	0.70	0.34	0.36	0.32	—	0.77	0.74	0.28	0.37	0.28	—
3β-ol,17-one/17-one	0.73	0.71	0.37	0.42	—	—	0.81	0.75	-0.70	—	—	—
3α-ol												
3α,17β-diol/17β-ol	0.68	0.58	0.19	0.39	0.37	—	0.74	0.63	0.10	0.41	0.35	—
3α,ol,17-one/17-one	0.67	0.62	0.17	0.40	0.39	—	0.79	0.63	0.08	0.49	0.40	-0.05
17-one												
17-one/androstane	0.64	—	—	0.94	—	—	0.68	—	—	0.98	—	—
3β-ol,17-one/3β-ol	0.65	0.63	0.64	0.69	—	—	0.72	0.68	-0.30	—	—	—
3,17-dione/3-one	0.67	—	—	0.67	—	—	0.74	—	—	0.73	—	—
3-one												
3-one/androstane	0.73	—	—	0.72	—	—	0.76	—	—	0.74	—	—
3-one,17β-ol/17β-ol	0.74	0.73	0.73	0.42	0.39	—	0.81	0.82	0.76	0.45	0.43	—
3, 17-dione/17-one	0.76	—	—	0.44	—	—	0.82	—	—	0.50	—	—

*Reproduced by permission from I.S. Hartman and H.H. Wotiz, Biochim. Biophys. Acta 90:334-348 (1964), Table III.

TABLE 13-11.   Retention Factors for 3β-Methoxyl and 5α-Hydroxyl Groups*

Hydrocarbon	Relative retention			Retention factors	
	Unsubstituted	With substituent groups		$k_{3\beta\text{-}OMe}$	$k_{5\,\alpha\text{-}OH}$
		3β-Methoxyl	5α-Hydroxyl		
Androstane	0.12	0.42		3.5	
Allopregnane	0.21	0.73	1.10	3.5	5.3
Cholestane	1.00	3.60	5.24	3.6	5.2
Δ^2-Cholestene	1.20		5.34		4.4
Δ^8-Cholestene	1.10	3.88		3.5	
Ergostane	1.36	4.81	6.98	3.5	5.1
Δ^7-Ergostene	1.78	6.05	7.26	3.4	4.1
$\Delta^{8(14)}$-Ergostene	1.40	5.10		3.6	
$\Delta^{7,22}$-Ergostadiene	1.57	5.55		3.5	
Stigmastane	1.72	6.10	8.70	3.6	5.1

Observed and Calculated† Relative Retentions of Some Mono- and Bifunctional Steroids

Monofunctional steroids	Relative retention	Bifunctional steroids	Relative retention
17β-Methoxyandrostane	0.333	3β,17β-Dimethoxyandrostane	1.19 (1.15)
7α-Methoxycholestane	1.69	3β,7α-Dimethoxycholestane	5.90 (6.10)
7β-Methoxycholestane	2.30	3β,7β-Dimethoxycholestane	7.90 (8.30)
		3β,7α-Dimethoxyallopregnane	1.19 (1.22)
		3β,7β-Dimethoxyallopregnane	1.60 (1.68)
7α-Hydroxycholestane	6.83	3β-Methoxy-7α-hydroxyallopregnane	5.20 (5.00)

Relative Retentions of Some Bile Acid Derivatives

Compound	Relative retention	$k\,\Delta^7$	$k\,\Delta^9$	$k\,\Delta^{7,9}$	$k_{3\alpha\text{-}OMe}$
Methyl cholanate	3.81				
3α-Methoxymethyl cholanate	12.8				3.40
Methyl Δ^7-cholenate	4.31	1.13			
3α-Methoxymethyl Δ^7-cholenate	14.85	1.16			3.45
Methyl Δ^9-cholenate	3.41		0.90		
3α-Methoxymethyl Δ^9-cholenate	11.85		0.93		3.48
3α-Methoxymethyl $\Delta^{7,9}$-choladienate	14.0			1.09	

*Reproduced by permission from R. B. Clayton, Biochemistry 1:357–366 (1962), Table VIII.
†Calculated values in parentheses.

TABLE 13-12. Regularities in Separation Factors for 5α- and 5β-Steroids on SE-30 and QF-1 Columns at 200°[*]

5α-and 5β-Epimers	Separation factor[†]	
	SE-30	QF-1
1. Androstane	1.11	—
2. Pregnane	1.11	1.14
3. Pregnan-20-one	1.11	1.11
4. Cholane	1.10	1.10
5. Cholan-12-one	1.11	1.19
6. Cholan-12α-ol	1.14	1.17
7. Cholestane	1.11	1.11
8. Androstane-3,17-dione	1.12	1.08
9. 17β-Hydroxyandrostan-3-one	1.12	1.09
10. Androstane-3,11,17-trione	1.16	1.12
11. Pregnane-3,20-dione	1.12	1.09
12. Pregnane-3,11,20-trione	1.15	1.15
13. Cholestan-3-one	1.11	1.11
14. 3α-Hydroxyandrostan-17-one	1.10	0.95
15. Pregnan-3α-ol	1.07	0.99
16. Cholestan-3α-ol	1.11	1.00
17. Pregnan-3β-ol	1.11	1.28
18. 3β-Hydroxypregnan-20-one	1.13	1.26
19. Cholestan-3β-ol	1.13	1.29
20. 3α-Hydroxyandrostan-17-one 3-acetate[‡]	1.02	1.06
21. Pregnan-3α-ol 3-acetate[‡]	1.00	1.08
22. Cholestan-3α-ol 3-acetate[‡]	1.00	1.08
23. Pregnan-3β-ol 3-acetate[‡]	1.23	1.18
24. 3β-Hydroxypregnan-20-one 3-acetate[‡]	1.23	1.17
25. Cholestan-3β-ol 3-acetate[‡]	1.22	1.17

[*]Reproduced by permission from C. J. W. Brooks and L. Hanaineh, Biochem. J. 87:151-161 (1963), Table II.
[†]The separation factor is the ratio of the retention of the 5α-steroid to that of its 5β-epimer.
[‡]Compounds 20–25 were prepared by acetylation of compounds 14–19 and were not isolated; the relative retention data are in Table 13-8

TABLE 13-13. C-5 Isomeric Separation Factors*

Steroid	SE-30 3%			XE-60 3%		
	Free	Acetate	TMSi	Free	Acetate	TMSi
5α-Androstane/5β-androstane	1.04	–	–	1.08	–	–
5α-Androstan-17β-ol/5β-androstan-17β-ol	0.80	0.76	–	0.70	0.68	0.73
5α-Androstan-17-one/5β-androstan-17-one	0.76	–	–	0.68	–	–
5α-Androstan-3β-ol/5β-androstan-3β-ol	1.09	1.11	–	1.23	1.17	1.19
5α-Androstan-3-one/5β-androstan-3-one	1.03	–	–	1.13	–	–
5α-Androstane-3α,17β-diol/5β-androstane-3α,17β-diol	1.04	1.02	0.95	0.99	1.05	0.98
5α-Androstane-3β,17β-diol/5β-androstane-3β,17β-diol	1.19	1.14	1.22	1.21	1.16	1.29
5α-Androstan-3α-ol,17-one/5β-androstan-3α-ol,17-one	1.10	1.03	1.02	1.05	1.00	0.93
5α-Androstan-3β-ol,17-one/5β-androstan-3β-ol,17-one	–	–	–	1.06	0.82	1.35
5α-Androstan-3α-ol,17-one/androst-5-en-3β-ol,17-one	0.85	0.90	0.86	1.00	0.90	0.79
5β-Androstan-3α-ol,17-one/androst-5-en-3β-ol,17-one	0.77	0.87	0.84	0.95	0.90	0.85
5α-Androstan-3β-ol,17-one/androst-5-en-3β-ol,17-one	0.91	1.02	1.00	1.04	1.03	1.05
5β-Androstan-3β-ol,17-one/androst-5-en-3β-ol,17-one	–	–	–	0.99	1.25	0.78
5α-Androstan-3-on,17β-ol/5β-androstan-3-on,17β-ol	1.08	1.10	1.08	1.06	1.11	1.12
5α-Androstan-3-on,17α-ol/5β-androstan-3-on,17α-ol	1.06	1.23	1.26	0.98	1.08	1.28
5α-Androstane-3,17-dione/5β-androstane-3,17-dione	–	–	–	–	–	–

TABLE 13-13 cont.

Steroid	NGSeb 3%			Hi-Eff 8B 3%		
	Free	Acetate	TMSi	Free	Acetate	TMSi
5α-Androstane/5β-androstane	1.13	—	—	1.06	—	—
5α-Androstan-17β-ol/5β-androstan-17β-ol	0.56	0.55	—	0.48	0.46	—
5α-Androstan-17-one/5β-androstan-17-one	0.55	—	—	0.53	—	—
5α-Androstan-3β-ol/5β-androstan-3β-ol	1.24	1.27	—	1.23	1.25	—
5α-Androstan-3-one/5β-androstan-3-one	1.17	—	—	1.09	—	—
5α-Androstane-3α,17β-diol/5β-androstane-3α,17β-diol	1.09	0.89	0.89	1.02	0.88	0.76
5α-Androstane-3β,17β-diol/5β-androstane-3β,17β-diol	1.26	1.33	1.57	1.20	1.33	1.50
5α-Androstan-3α-ol,17-one/5β-androstan-3α-ol,17-one	1.02	0.91	0.81	1.04	0.92	0.72
5α-Androstan-3β-ol,17-one/5β-androstan-3β-ol,17-one	1.12	—	—	—	—	—
5α-Androstan-3α-ol,17-one/androst-5-en-3β-ol,17-one	0.88	0.81	0.64	0.93	0.78	0.64
5β-Androstan-3α-ol,17-one/androst-5-en-3β-ol,17-one	0.86	0.88	0.80	0.90	0.85	0.89
5α-Androstan-3β-ol,17-one/androst-5-en-3β-ol,17-one	1.00	1.01	1.00	0.99	1.02	1.03
5β-Androstan-3β-ol,17-one/androst-5-en-3β-ol,17-one	0.90	—	—	—	—′	—
5α-Androstan-3-on,17β-ol/5β-androstan-3-on,17β-ol	1.19	1.20	1.18	1.10	1.13	1.12
5α-Androstan-3-on,17α-ol/5β-androstan-3-on,17α-ol	1.08	1.20	1.52	1.10	1.14	1.35
5α-Androstane-3,17-dione/5β-androstane-3,17-dione	1.15	—	—	1.10	—	—

*Reproduced by permission from I.S. Hartman and H.H. Wotiz, Biochim. Biophys. Acta 90:334–348 (1964), Table VIII.

TABLE 13-14. Observed and Calculated Relative Retentions (r) of Some Mono- and Bifunctional Steroids*

Monofunctional steroids	r (obs.)	Bifunctional steroids	r (obs.)	r (calc.)
17β-Methoxyandrostane	0.333	3β,17β-Dimethoxyandrostane	1.19	1.18
7α-Methoxycholestane	1.67	3β,7α-Dimethoxycholestane	5.65	5.83
7β-Methoxycholestane	2.25	3β,7β-Dimethoxycholestane	7.65	7.88
		3β,7α-Dimethoxyallopregnane	1.19	1.22
		3β,7β-Dimethoxyallopregnane	1.60	1.64
7α-Hydroxycholestane	6.83	3β-Methoxy-7α-hydroxyallopregnane	5.20	5.00
Δ^2-Cholestene	1.20	5α-Hydroxy-Δ^2-cholestene	5.34	6.30
Δ^7-Ergostene	1.80	5α-Hydroxy-Δ^7-ergostene	7.26	9.06

*Reproduced by permission from R. B. Clayton, Biochemistry 1:357–366 (1962), Table II.

TABLE 13-15. Comparison of Calculated* and Observed Retention Values†

Steroids	3% SE-30		3% XE-60		3% Hi-Eff 8B		3% NGSeb	
	Calc.	Obs.	Calc.	Obs.	Calc.	Obs.	Calc.	Obs.
Free:								
5α-Androstane-3,17-dione	0.60	0.58	2.88	2.64	4.17	4.00	2.19	2.27
5β-Androstane-3,17-dione	0.50	0.50	2.52	2.48	3.17	3.64	1.59	1.98
5α-Androstan-3-on, 17β-ol	0.63	0.59	2.58	2.58	4.68	4.61	2.57	2.61
5β-Androstan-3-on, 17β-ol	0.50	0.54	2.24	2.44	3.32	4.18	1.74	2.20
Acetate:								
5α-Androstane-3β,17β-diol	1.00	0.98	2.75	2.21	3.39	3.18	2.19	2.20
5β-Androstane-3β,17β-diol	0.81	0.86	2.24	1.90	—	2.42	—	1.66
TMSi ether:								
5α-Androstane-3β,17β-diol	0.69	0.69	0.71	0.66	—	0.42	0.40	0.44
5β-Androstane-3β,17β-diol	—	—	0.47	0.51	—	0.28	—	0.28

*Based on average log k values.
†Reproduced by permission from I. S. Hartman and H. H. Wotiz, Biochim. Biophys. Acta 90:334–348 (1964), Table IV.

TABLE 13-16. Log r (trimeth-
ylsilylation) Values on 3% NGA[*]
at 225°

Group reacted	Log r
3-OH (equatorial, 5α)	0.41
3-OH (equatorial, 5β)	0.44
3-OH (axial, 5α)	0.55
3-OH (axial, 5β)	0.62

[*]Neopentylglycol adipate.

TABLE 13-17. Retention Factors to Acetylation of 3-Hydroxy
Steroids[*]

	SE-30			QF-1		
	Relative retention data		Factor due to acetyl-ation	Relative retention data		Factor due to acetyl-ation
	Hydroxy steroid	Acetoxy steroid		Hydroxy steroid	Acetoxy steroid	
Axial: 3α-substituted 5α-steroid						
Androstan-17-one	0.336	0.474	1.41	1.957	3.270	1.67
Pregnane	0.304	0.436	1.43	0.647	1.034	1.60
Cholestane	2.120	3.089	1.46	3.237	5.18	1.60
Axial: 3β-substituted 5β-steroid						
Pregnane	0.279	0.406	1.46	0.574	0.994	1.73
Pregnan-20-one	0.523	0.769	1.47	2.486	4.42	1.78
Cholestane	1.892	2.806	1.48	2.829	4.90	1.73
Equatorial: 3β-substituted 5α-steroid						
Pregnane	0.311	0.499	1.61	0.733	1.118	1.60
Pregnan-20-one	0.593	0.943	1.50	3.143	5.16	1.64
Cholestane	2.135	3.411	1.60	3.662	5.718	1.56
Equatorial: 3α-substituted 5β-steroid						
Androstan-17-one	0.306	0.467	1.53	2.051	3.083	1.50
Pregnane	0.283	0.437	1.54	0.653	0.961	1.47
Cholestane	1.918	3.077	1.60	3.240	4.81	1.48

[*]The acetates were prepared from 50 μg. of hydroxy steroid and were not isolated. The
columns were operated at 200°.
Reproduced by permission from C. J. W. Brooks and L. Hanaineh, Biochem. J. 87:151 (1963).

time values as compared to the retention times of the free compounds using the equation

$$r_{derivative}/r_{free} \times 100 = r'\%$$

Interactions of the relatively nonpolar acetyl groups with the nonpolar SE-30 and XE-60 liquid phases are indicated in the increased retention times of the steroid acetates on these columns as shown in Table 13-18.

The decreased interactions of the acetates with the polar phases are reflected in the faster elution time of these derivatives as compared to the free steroid on the Hi-Eff 8B and the neopentylglycol sebacate columns. It is also evident, from the data presented in Table 13-18, that of the three derivatives, the contribution to retention time is additive only in the case of the acetate group. For example, on the XE-60 column, there is a 10% increase in retention time upon substitution of one acetate group and an additional 10% increase (a total of 20%) upon the substitution of a second acetate group. The absence of a direct relationship between the number of substituent groups and their contribution to retention time in the case of trimethylsilyl ethers may be indicative of the presence of factors such as steric hindrance. The use of acetates in preference to trimethylsilyl derivatives in structural determination studies have therefore been suggested by Hartman and Wotiz [10]. Apart from being useful in the characterization of functional groups, a knowledge of r values of closely related steroids might be useful in the interpretation of gas chromatograms of complex mixtures of steroids. Relative retentions may vary slightly, but once relative orders of elution for steroid epimers have been established, complex mixtures may be analyzed by r values with reference to any one epimer identified in the mixture or chromatographed at the same time. Such r parameters might also be used to predict the chromatographic properties of compounds not readily available for use as reference standards.

STEROID NUMBERS AND T VALUES

The approach described in the previous section is based on the use of different stationary phases to confirm or predict the results obtained on any one phase. A second, though less well-developed, approach is the correlation of change in retention time with chemical structure due to a change in stationary phase. VandenHeuvel and Horning [18–19a] have suggested two new terms, steroid number (SN) and T, as parameters for the elucidation of the structural feature of a steroid molecule.

Both of these terms are directly derived from the work of Woodford and van Gent [20] in the gas–liquid chromatography of fatty acids.

Steroid number is analogous to carbon number of fatty acids, which is related to retention by

$$\log r = k \times \text{number of carbon atoms in fatty acid molecule}$$

The carbon number may be determined by constructing a reference graph according to the above equation, based on the carbon number of known fatty acids. The carbon number of an unknown can then be read off at a point corresponding to the retention time of the acid. The concept of carbon number may be useful because it is independent of slight changes in temperature and eliminates the need for an arbitrary reference standard to which retention times must be related.

Steroid number is expressed as a summation of terms dependent on the nature of the carbon skeleton and the functional groups on the steroid molecule

$$SN = S + F_a + F_b + \ldots$$

where SN is the steroid number, S is the carbon content of the steroid skelton, and F_a and F_b are values characteristic of the functional groups of the steroids. The SN value of a particular steroid is found by comparing the retention times of a steroid with that of cholestane (steroid number 27) and androstane (steroid number 19) under the same conditions. A nonselective stationary phase such as SE-30 is desirable for use in determining this parameter since nonselective phase separations are based on the molecular-weight differences of the samples under investigation [10, 17]. Tables 13-19 and 13-20 list the SN values of some representative steroids as determined by VandenHeuvel and Horning. Figure 13-5 shows the graph used to calculate these values.

VandenHeuvel and Horning have found steroid numbers to be useful in defining relationships based primarily on molecular weight and molecular shape, ascribing, for example, the difference in steroid

TABLE 13-18. The Average Change in Retention Time Occurring in Derivative Formation ($r_{deriv}/r_{free} \times 100 = r'$, in percent)[*]

Column	Acetate		TMSi		TFA	
	Mono	Di	Mono	Di	Mono	Di
SE-30 3%	123	169	102	115	69	59
XE-60 3%	110	120	53	30	–	–
Hi-Eff 8B 3%	83	70	27	9	23	7
NGSeb 3%	93	85	41	16	40	15

[*]Reproduced by permission from I. S. Hartman and H. H. Wotiz, Biochim. Biophys. Acta 90:334-348 (1964), Table VI.

TABLE 13-19. Steroid Numbers and Relative Retentions for Representative Steroids Determined at 211° with an SE-30 Phase[*]

Steroid	Retention relative to cholestane	Steroid number (SN)
Androstan-17-one	0.18	21.3
Androstane-3,17-dione	0.39	23.9
4-Androstene-3,17-dione	0.48	24.6
4-Androsten-17β-ol-3-one	0.49	24.7
Androstane-3,16-dione	0.40	23.9
4-Androstene-3,16-dione	0.48	24.6
4-Androstene-3,11,17-trione	0.57	25.1
4-Androstene-11α-ol-3,17-dione	0.85	26.5
4-Androstene-11β-ol-3,17-dione	0.81	26.3
19-Nor-4-androstene-3,17-dione	0.38	23.8
Androstane-3β,17β-diol	0.38	23.8
Androstane-3α,17β-diol	0.37	23.7
Androstan-3β-ol-17-one	0.37	23.7
Androstan-3α-ol-17-one	0.36	23.6
5-Androstene-3β,17β-diol	0.37	23.7
5-Androsten-3β-ol-17-one	0.36	23.6
17α-Methyl-5-androstene-3β,17β-diol	0.41	24.1
Androstane-3β,16β-diol	0.37	23.7
Estrone	0.46	24.4
Estradiol	0.48	24.6
Equilenin	0.61	25.4
Pregnane-3,20-dione	0.60	25.3
Pregnan-21-ol-3,20-dione	1.10	27.4
5α-Pregnane-3,20-dione	0.66	25.6
5α-Pregnane-3,11,20-trione	0.89	26.7
5α-Pregnane-3β,20α-diol	0.66	25.6
5α-Pregnane-3β,20β-diol	0.62	25.4
5α-Pregnan-20β-ol-3-one	0.66	25.6
5α-Pregnan-3β-ol-20-one	0.62	25.4
5α-Pregnan-3β-ol-11,20-dione	0.86	26.6
4-Pregnene-3,20-dione	0.78	26.2
Cholestanol	2.00	29.4
Epicholestanol	1.98	29.4
Coprostanol	1.80	29.1
Cholestan-3-one	2.14	29.6
4-Cholesten-3-one	2.64	30.3
Cholestanyl acetate	3.12	30.9
Cholestanyl methyl ether	1.88	29.2
Cholestanyl trifluoracetate	1.66	28.8
Cholestanyl trimethylsilyl ether	2.61	30.3
Tigogenin	2.71	30.7
Solanidan-3β-ol	2.24	29.8
7-Cholesten-3β-ol	2.21	29.8
Desmosterol	2.12	29.6
Cholesterol	1.95	29.3

[*]Reproduced by permission from W. J. A. VandenHeuvel and E. C. Horning, Biochim. Biophys. Acta 64:416-429 (1962), Table I. Conditions: 6 ft by 4 mm glass U-tube; 1% SE-30 on 100-120 mesh Gas Chrom P, at 211°, 20 lb/in.^{2}, cholestane time 13.5 min.

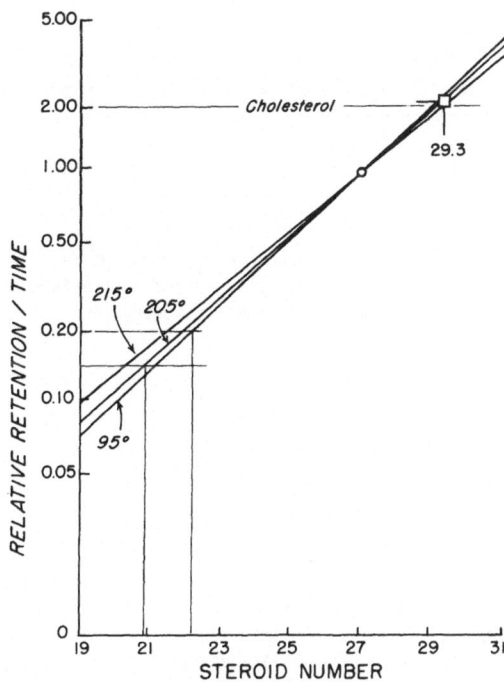

Figure 13-5. Relationship between temperature and the slope of the androstane–cholestane line used to determine steroid numbers from relative retention times [W. J. A. VandenHeuvel and E. C. Horning, Biochem. Acta. 64:416-429, (1962)].

number of the 3α-trimethylsilyloxy group (axial, SN 2.5) and the 3β-trimethylsilyloxy group (equatorial SN 3.3) to the relatively great differences in the molecular shape of these steroids. However, the absence of such a difference between the steroid number values of 3α-OH and 3β-OH groups is not explained. It should be noted that the advantage, reliability, and usefulness of the SN values in the elucidation of structure of steroids are still questionable for the following reasons. First, Woodford and van Gent employed a standard curve based on the chromatographic behavior of five fatty acids of known chain length. VandenHeuvel and Horning have based their calculations on a standard curve made up of two points. Reliability of any values obtained from such a curve may be questioned. Secondly, the steroid number term neither reflects clearly the actual carbon content of the steroid molecule nor indicates substituent groups on the nucleus with any accuracy. For example, the values of SN as seen from Table 13-18 are 29.4 for cholestanol, 29.2 for cholestanyl methyl ether, and 29.4 for cholestanyl trifluoroacetate. Neither the size nor the shape of the molecule is clearly described by these values.

The second parameter T , again derived from the earlier work

TABLE 13-20.   Steroid Number Contributions for Representative Functional Groups Determined with a Nonselective Phase (SE-30)[*]

Functional group	Steroid parent[†]	Steroid number (F)[‡]
Δ^5	(Cholestanol/cholesterol)	0.1
Δ^5	(Cholestane/Δ^5-cholestene)	0.0
Δ^7	(Cholestanol/7-cholesten-3β-ol)	0.4
Δ^{24}	(Cholesterol/desmosterol)	0.3
3-one	(Androstane)	2.6
3-one	(5α-Pregnane)	2.6
3-one	(Cholestane)	2.6
3-one-Δ^4	(Androstane)	3.3
3-one-Δ^4	(Cholestane)	3.3
11-one	(Δ^4-Androstene)	0.5
11-one	(5α-Pregnane)	1.1
16-one	(Androstane)	2.3
17-one	(Androstane)	2.3
20-one	(5α-Pregnane)	2.0
3α-ol (axial)	(Androstane)	2.3
3β-ol (equatorial)	(Androstane)	2.4
3β-ol (equatorial)	(5α-Pregnane)	2.4
3β-ol (equatorial)	(Cholestane)	2.4
11α-ol (equatorial)	(Androstane)	1.9
11β-ol (axial)	(Androstane)	1.7
16β-ol	(Androstane)	2.3
17β-ol (secondary)	(Androstane)	2.4
17β-ol (tertiary)	(17α-Methylandrostane)	1.7
20α-ol	(5α-Pregnane)	2.2
20β-ol	(5α-Pregnane)	2.0
21-ol	(5α-Pregnane)	2.1
Aromatic B ring	(Estrone/equilenin)	1.0
N	(Solanidan-3β-ol)	0.4
Spiroketal	(Tigogenin)	1.3
3β-Trifluoracetoxy (equatorial)	(Cholestane)	1.8
3β-Methoxy (equatorial)	(Cholestane)	2.2
3α-Trimethylsilyloxy (axial)	(Cholestane)	2.5
3β-Trimethylsilyloxy (equatorial)	(Cholestane)	3.3
3β-Acetoxy (equatorial)	(Cholestane)	3.9
A/B cis	(Coprostane/cholestane)	−0.2
A/B cis	(Coprostanol/cholestanol)	−0.3

[*]Reproduced by permission from W. J. A. VandenHeuvel and E. C. Horning, Biochim. Biophys. Acta 64:416–429 (1962), Table II.
[†]Structure of the steroid nucleus for the reference compound used in the determination of the SN value.
[‡]Determined with the column and conditions described in Table 13-8.

TABLE 13-21. Values of T Determined at 211° for Representative
Functional Groups with NGS and QF-1 Phases*

Functional group	Steroid	T values	
		NGS	QF-1
3-one	Cholestan-3-one	2.1	2.0
3-one-Δ^4	4-Cholesten-3-one	2.9	2.9
17-one	Androstan-17-one	2.5	2.6
3α-ol (axial)	Cholestan-3α-ol	1.9	0.4
3β-ol (equatorial)	Cholestan-3β-ol	2.2	0.7
3β-ol-Δ^5 (equatorial)	Cholesterol	2.5	0.6
3, 17-dione	Androstane-3, 17-dione	12.1	10.7
3, 17-dione-Δ^4	4-Androstene-3, 17-dione	15.6	13.5
3, 20-dione (A/B cis)	Pregnane-3, 20-dione	9.7	8.2
3, 20-dione (A/B trans)	5α-Pregnane-3, 20-dione	9.9	7.8
3β, 17β-diol	Androstane-3β, 17β-diol	11.8	2.5
3α, 17β-diol	Androstane-3α, 17β-diol	10.7	2.3
3β, 17β-diol-Δ^5	5-Androstene-3β, 17β-diol	12.5	2.3
3β, 20β-diol	5α-Pregnane-3β, 20β-diol	9.4	2.1
3β, 20α-diol	5α-Pregnane-3α, 20α-diol	10.6	2.5
3β-ol-17-one	Androstan-3β-ol-17-one	11.0	5.1
3α-ol-17-one	Androstan-3α-ol-17-one	10.0	4.4
3β-ol-17-one-Δ^5	5-Androsten-3β-ol-17-one	12.1	4.5
3β-ol-20-one	5α-Pregnan-3β-ol-20-one	9.5	3.8
17β-ol-3-one-Δ^4	4-Androsten-17β-ol-3-one	16.5	7.1
20β-ol-3-one	5α-Pregnan-20β-ol-3-one	10.6	4.6
3, 17-dione-Δ^4-11β-ol	4-Androsten-11β-ol-3, 17-dione	42.0	15.0
3, 11, 20-trione (A/B trans)	5α-Pregnane-3, 11, 20-trione	—	15.1
3-ol-17-one, A aromatic	Estrone	34.0	4.9
3-ol-17β-ol, A aromatic	Estradiol	37.2	2.5
3α-Trimethylsilyloxy (axial)	Epicholestanyl TMSi	−0.4	−0.2
3β-Trimethylsilyloxy (equatorial)	Cholestanyl TMSi	−0.2	−0.1
3β-Trifluoracetoxy (equatorial)	Cholestanyl TFA	0.3	0.8
3β-Trifluoracetoxy-Δ^5 (equatorial)	Cholesteryl TFA	0.3	0.7
3β-Methoxy (equatorial)	Cholestanyl ME	0.4	0.2
3β-Acetoxy (equatorial)	Cholestanyl acetate	0.8	0.7
A/B cis	5α- and Pregnane-3, 20-dione	−0.2	0.4

*Reproduced by permission from W. J. A. VandenHeuvel and E. C. Horning, Biochim. Biophys.
Acta 64:416-429 (1962), Table VI.

of Woodford and van Gent [20], is based on the measurement of relative retentions on two stationary phases and is thus a function of the degree of selectivity of the stationary phase. The term is defined as

$$T = \frac{t'_s - t'_n}{t'_n}$$

where t'_s is the relative retention observed with a selective stationary phase, and t'_n is the relative retention observed with a nonselective stationary phase. Woodford and van Gent had employed polar and nonpolar columns to study the behavior of saturated fatty acids in an analogous manner. Some representative T values are given in Table 13-21.

When the phase is neopentylglycol succinate polyester, it is seen that the T values closely follow the polarity of the functional groups. With QF-1, T values clearly reflect the ketone selectivity of this phase. T values of androstane isomers (Table 13-5) calculated from measurements obtained on the XE-60 and SE-30 phases indicate the dominating influence of the C-17 oxygen function in the 5β-series when the C-3 hydroxyl group is in the alpha configuration. Similarly, the ketone selective properties of the XE-60 phase are observable from these data. Thus, T values might be useful in reflecting the number, nature, and stereochemical relationships of functional groups and might provide clues to intramolecular interactions. However, further work is necessary before either SN or T values can be used to derive a meaningful, if tentative, conclusion about the structure of an unidentified steroid molecule.

REFERENCES

1. A. J. P. Martin, Biochem. Soc. Symp. 3:4 (1949).
2. A. England and E. S. Cohn, J. Am. Chem. Soc. 57:634 (1935).
3. E. C. Bate-Smith and R. G. Westall, Biochim. Biophys. Acta 4:427 (1950).
4. I. E. Bush, The Chromatography of Steroids (Pergamon Press, New York, 1961), Chapt. 1 and 2.
5. R. B. Clayton, Nature 190:1071 (1961).
6. R. B. Clayton, Nature 192:524 (1961).
7. R. B. Clayton, Biochemistry 1:357 (1962).
8. B. A. Knights and G. H. Thomas, Anal. Chem. 34:1046 (1962).
8a. B. A. Knights and G. H. Thomas, J. Chem. Soc. 3477 (1963).
8b. B. A. Knights and G. H. Thomas, Nature 194:833 (1962).
9. I. S. Hartman and H. H. Wotiz, Steroids 1:33 (1963).
10. I. S. Hartman and H. H. Wotiz, Biochim. Biophys. Acta 90:334 (1964).
11. K. Savard, J. Biol. Chem. 202:457 (1953).
12. D. H. R. Barton, Experientia 6:316 (1950).
13. D. H. R. Barton and W. J. Rosenfelder, J. Chem. Soc. 1048 (1951).
14. C. J. W. Brooks and L. Hanaineh, Biochem. J. 87:151 (1963).
15. S. R. Lipsky and R. A. Landowne, Anal. Chem. 33:818 (1961).
16. B. A. Knights and G. H. Thomas, Chem. Ind. 43 (1963).
17. H. H. Wotiz and H. F. Martin, J. Biol. Chem. 236:1312 (1961).
18. W. J. A. VandenHeuvel and E. C. Horning, Biochim. Biophys. Acta 64:416 (1962).
19. E. O. A. Haahti, W. J. A. VandenHeuvel, and E. C. Horning, Biochim. Biophys. Acta 70:679 (1963).
19a. R. J. Hamilton, W. J. A. VandenHeuvel, and E. C. Horning, Biochim. Biophys. Acta 70:679 (1963).
20. F. P. Woodford and C. M. van Gent, J. Lipid Res. 1:188 (1960).

Chapter 14

Determination of Progesterone
and Its Metabolic Products

BIOCHEMISTRY

Progestational changes of the endometrium of the uterus can be induced by a number of steroid substances commonly referred to as progestins. The early work of Corner and Allen [1] showed that in the normal female a progestational hormone is released by the corpus luteum. Several research groups in the early 1930's proved this to be the steroid progesterone.

PROGESTERONE
(pregn-4-ene-3, 20-dione)

Although a number of synthetic substances containing the steroid nucleus are presently known which produce these progestational reactions, only progesterone has been isolated and identified as the primary biosynthetic product of the corpus luteum. Its reduction products at C-20 (20α-hydroxy-pregn-4-en-3-one and 20β-hydroxy-pregn-4-en-3-one) also have been identified as naturally occurring substances with progestational activity. In all likelihood, these are derived by reduction from the original progesterone of the corpus luteum.

After the biologic utilization of progesterone, several reaction products can be identified. Despite a number of investigations only a few of the metabolic products, amounting to approximately 25% of an administered dose of pure progesterone, have as yet been characterized. Quantitatively, the most significant metabolic product derived from progesterone is 5β-pregnane-3α, 20α-diol.

129

PREGNANEDIOL
(5β-pregnane-3α, 20α-diol)

Next in line appears to be a compound known as pregnanolone

PREGNANOLONE
(3α-hydroxy-5β-pregnan-20-one)

The presence of allopregnanediol has been demonstrated in pregnancy urine. A number of other reduction products are also

ALLOPREGNANEDIOL
(5α-pregnane-3α, 20α-diol)

likely metabolites; however, they are not found in any significant concentration compared to the other three substances. In view of the relative lack of knowledge regarding the total metabolic changes occurring following the utilization of progesterone, it is generally conceded that the best method of evaluating the function of the corpus luteum or the placenta is based on the determination of urinary pregnanediol and/or the measurement of plasma progesterone.

Since the adrenal gland is also a source of progesterone, the measurement of pregnanediol to evaluate adrenal function has been advocated by Klopper [2]. Further evidence for this will be presented later in the chapter.

DETERMINATION OF URINARY PREGNANEDIOL

Evaluation of Methods

A survey of the literature leads to the fair summation that the most acceptable and useful conventional method for the determination of urinary pregnanediol is the one described by Klopper et al. [3]. This method incorporates a simple acid hydrolysis and simultaneous toluene extraction, removal of the acidic fraction, oxidative purification followed by adsorption chromatography, acetylation, and further chromatography, and finally quantitation using the sulfuric acid chromogen. The procedure has been used to determine levels of pregnanediol excretion of as low as 0.5 mg per 24 hr and has been accepted as a method with a high degree of specificity. Despite the elegance of this procedure, application in a clinical laboratory suffers from one drawback—namely, the length of time involved between obtaining the specimen and the availability of the test result.

The application of gas chromatography to the determination of pregnanediol has two obvious advantages. The first is the significant increase in speed of the determination itself, and the second is the reduction in the number of manipulative procedures during the assay, allowing a greater recovery of pregnanediol. As pointed out in Chapter 6 of this book, a number of independent investigators have described, rather briefly, some methods for the determination of pregnanediol. Thus far, with the exception of two reports [4, 5], most of these have dealt only in a preliminary manner with the subject and it is as yet hard to judge them individually. Two major questions should be answered in surveying all these procedures. The first relates to the choice of derivative, if any, and the second to the necessity of extensive preliminary cleanup.

The clearest evidence that derivative formation is necessary is reported by Kirschner and Lipsett [4] who have found, as we have in our laboratory, that there is one unidentified urinary component which has a retention time very close to that of pregnanediol and which is not separated from the latter even after thin-layer chromatography. Neither the trimethylsilyl ethers nor the trifluoroacetates of the two substances allow their resolution; only when acetylation was effected was separation achieved.

The problem of whether it becomes necessary to introduce a preliminary cleanup, such as TLC, prior to analysis is not quite so easily disposed of. It depends upon a number of requisites which must be determined by the individual investigator. Preliminary purification is a desirable feature provided that sufficient time is available and benefit is derived from such a procedure.

Kirschner and Lipsett have accumulated a valuable set of numbers as evidence for the validity of the method. Specificity was determined

primarily on the basis of chromatographic procedures, namely by the
R_f values (Table 14-1) obtained in thin-layer chromatography and by
the GLC relative retention data (Table 14-2).   Unfortunately, no
statement is made as to whether retention time comparison of the
peaks from the extracts and the authentic steroids was made on an
SE-30 column only or on other stationary phases. Final evidence was
obtained by the addition of authentic material to extracts, with the
appearance of single peaks following further gas chromatography. The
sensitivity of the method compares to that reported by us (50 μg per
24 hr specimen).   At a level of excretion of 1.9 mg per 24 hr, a
coefficient of variation of 5% was obtained by Kirschner and Lipsett.
Recovery of material added in concentrations of 500 μg to 24 hr urines
was shown to be excellent.   It is interesting, though surprising,
in view of the evidence of Wotiz [6] and Cox [7], that a comparison of
methods indicates values for GLC determination of about one-half those
of the Klopper assay. As shown in Tables 14-3 and 14-4, both in our
laboratory and in Cox's laboratory, no such discrepancies have been
observed except at the very low levels of excretion. Furthermore, it
should be pointed out that the mean values of pregnanediol excretion

TABLE 14-1.   Values for Various Steroids in a Benzene:Ethyl Acetate
System (40:60)*

Steroid	Trivial name	Chemical name	R_f
$C_{19}O_2$:	Etiocholanolone	3α-Hydroxy-5β-androstan-17-one	0.41
	Androsterone	3α-Hydroxy-5α-androstan-17-one	0.47
	Dehydroepiandrosterone	3β-Hydroxy-androst-5-en-17-one	0.47
	Androstenediol	Androst-5-ene-3β,17β-diol	0.39
$C_{19}O_3$:	11β-Hydroxyandrosterone	3α,11β-Dihydroxy-5α-androstan-17-one	0.30
	11β-Hydroxyetiocholanolone	3α,11β-Dihydroxy-5β-androstan-17-one	0.26
	11-Ketoetiocholanolone	3α-Hydroxy-5β-androstan-11,17-dione	0.28
$C_{21}O_2$:	Pregnanediol	5β-Pregnane-3α,20α-diol	0.29
	Allopregnanediol	5α-Pregnane-3α,20α-diol	0.33
	Pregnenediol	Pregn-5-ene-3β,20α-diol	0.33
	Pregnanolone	3α-Hydroxy-5β-pregnan-20-one	0.47
	Pregnenolone	3β-Hydroxy-pregn-5-en-20-one	0.47
$C_{21}O_3$:	Pregnanetriol	5β-Pregnane-3α,17α,20α-triol	0.12
	Pregnenetriol	Pregn-5-ene-3β,17α,20α-triol	0.28
	17α-Hydroxypregnanolone	3α,17α-Dihydroxy-5β-pregnan-20-one	0.29
	17α-Hydroxypregnenolone	3β,17α-Dihydroxy-pregn-5-en-20-one	0.37

*Reproduced by permission from M. A. Kirschner and M. B. Lipsett, Steroids 3:283 (1964).

TABLE 14-2.   Chromatographic Characteristics
of Steroid Acetates*

Trivial name	Thin-layer†	GLC‡
Androsterone acetate	0.39	0.53
Etiocholanolone acetate	0.39	0.53
Dehydroepiandrosterone acetate	0.39	0.55
Pregnanediol diacetate	0.71	1.32
Allopregnanediol diacetate	0.71	1.49
Pregnanolone acetate	0.62	0.83
Pregnenediol diacetate	0.64	1.44
Pregnenolone acetate	0.57	1.00
17α-Hydroxypregnanolone acetate	0.39	1.17
17α-Hydroxypregnenolone acetate	0.39	1.29
Pregnanetriol diacetate	0.43	1.92
20β-Hydroxyprogesterone acetate	0.44	1.23

*Reproduced by permission from M. A. Kirschner and M. B. Lipsett, Steroids
3(3):285 (1964).
†Thin-layer system, benzene:ethyl acetate (80:20).
‡Retentions relative to cholestane on 1% SE-30.

for males, and females in the proliferative and luteal phases of the
cycle (Table 14-5), as reported by Kirschner and obtained in our own
laboratories are very similar.

Perhaps the main determinant in the application of a more
extensive prepurification rests with the efficiency of the gas chromatographic column. In our experience, even relatively low efficiency
columns with only a few hundred theoretical plates (Figure 14-1) can
be used to determine pregnanediol in pregnancy urine with considerable
reliability. Only when concentrations of less than approximately a
milligram per twenty-four hours are encountered do the neighboring
peaks begin to interfere with accurate measurement of peak height
(Figure 14-2). On the other hand, on columns with approximately two-
to three-thousand theoretical plates ($\bar{H}$ = 1 to $\bar{H}$ = 0.67), such problems
are rarely noted and even during the period just prior to ovulation, a
clean sharp peak may be obtained which allows ready measurement
(Figure 14-3). The first of these columns was calculated to have approximately eight-hundred theoretical plates, while Figures 14-3 and
14-4 were obtained from a column calculated to have approximately
3400 theoretical plates. From Figures 14-1 and 14-4 it is apparent that
accurate measurement may be made in either case. Figures 14-2 and
14-3, on the other hand, represent the chromatograms obtained on
these same two columns but from an extract of urine obtained during
the proliferative phase of the cycle. Quantification from the chro-

TABLE 14-3. Comparison of Pregnanediol Analyses[*]

Phase	Pregnanediol excretion (μg)		$\Delta\mu$g GLC−Klopper
	By Klopper[†] method	By gas chromatograph	
Luteal	3700	3820	+ 120
Follicular	550	320	−230
Follicular	480	310	−170
Luteal	1900	1980	+ 80
Luteal	6200	6230	+ 30
Follicular	350	180	−170
Luteal	2300	2380	+ 80

[*]Reproduced by permission from H. H. Wotiz and P. J. Mozden, Human Ovulation, edited by C. S. Keefer (Little, Brown and Company, Inc., Boston, 1965), Chapt. 10, Table 5. Analyses made simultaneously.
[†]For these analyses the Klopper procedure was modified by replacing the acetylation using acetyl chloride with the use of acetic anhydride and pyridine, as described in the text.

matogram in Figure 14-2 would be considered insufficiently reliable at this low concentration.

It is of course rather early in the development of this type of procedure to have adequate statistical data for its complete validification; nevertheless, a brief discussion relating to the major points of validity, as postulated by Borth (sensitivity, specificity, accuracy, and precision), is at least partially available at this time. Figure 14-6 shows a plot of peak height vs. concentration over a range of 2×10^3 (0.0075 to 16μg). The point values represent the means of three injections, none of which had a variance greater than ±1.8%. Linear response of the flame detector to pregnanediol diacetate is thereby demonstrated.

TABLE 14-4. Quantitative Estimation of Urinary Pregnanediol[*]

Sample	Pregnanediol found (mg per 24 hr)	
	By gas chromatography	By sulfuric acid chromogens
Pregnancy urine 1	37	38
Pregnancy urine 2	28	30
Pregnancy urine 3	8.9	9.0
Pregnancy urine 4	7.5	8.3
Pregnancy urine 5	4.2	3.8
Follicular phase urine	0.5	0.8

[*]Reproduced by permission from R. I. Cox, J. Chromatog. 12:242 (1963), Table II.

TABLE 14-5. Urinary Pregnanediol Excretion as Determined by GLC in Two Different Laboratories*

Sample	Mean excretion values (mg per 24 hr)		
	Kirschner and Lipsett	Wotiz	No. of analyses*
Males	0.41	0.46	9
Female, proliferative phase	0.55	0.52	100
Female, luteal phase	4.00	4.00	86

*Applies to center column only.

In order to study the specificity of the assay, a number of steroids similar in structure and related to pregnanediol were acetylated and chromatographed on an SE-30 column. Table 14-6 lists the retention relative to pregnanediol for some of these compounds. As expected, the acyl derivatives of the monohydroxy ketones appear much earlier in the chromatogram because of their lower molecular weight and present no difficulties in separation. Of the eight saturated isomeric diols, five were available to us; their retention times indicate that all

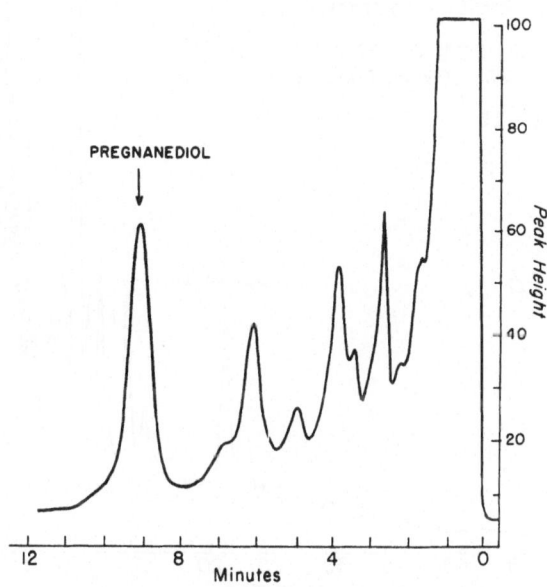

Figure 14-1. Gas chromatogram of a neutral pregnancy urine extract on an SE-30 column; Number of theoretical plates (n) = 1130.

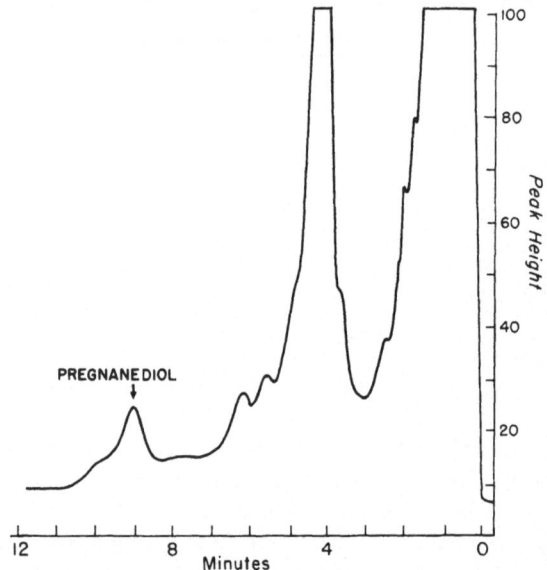

Figure 14-2. Gas chromatogram of a neutral proliferative phase urine extract on an SE-30
column; *n* = 1130.

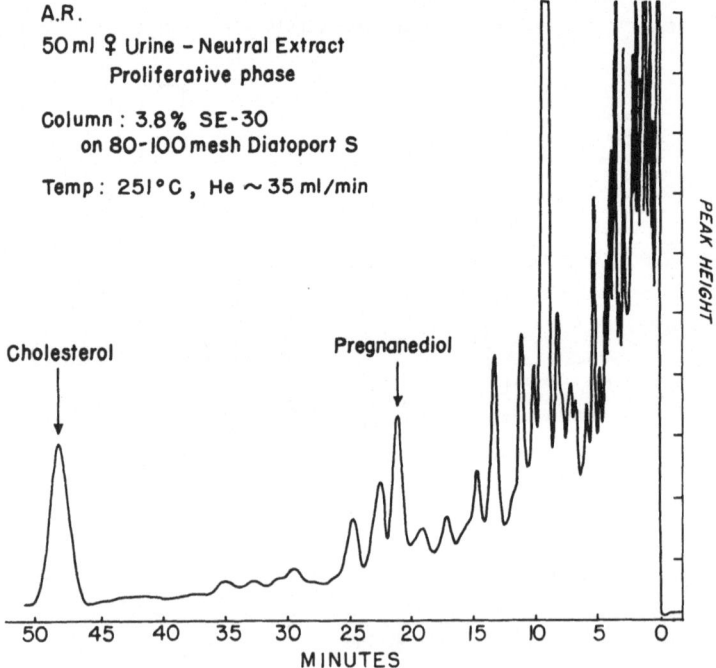

Figure 14-3. Gas chromatogram of a neutral proliferative phase urine extract on an SE-30
column; *n* = 4100.

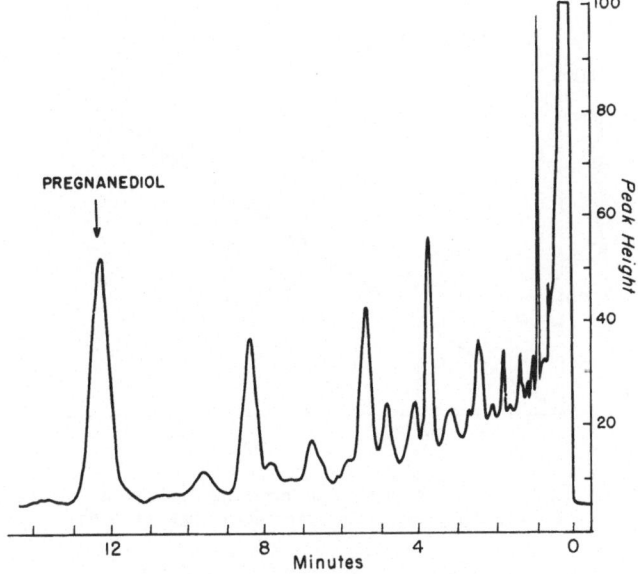

Figure 14-4. Gas chromatogram of a neutral pregnancy urine extract on an SE-30 column; $n = 3440$.

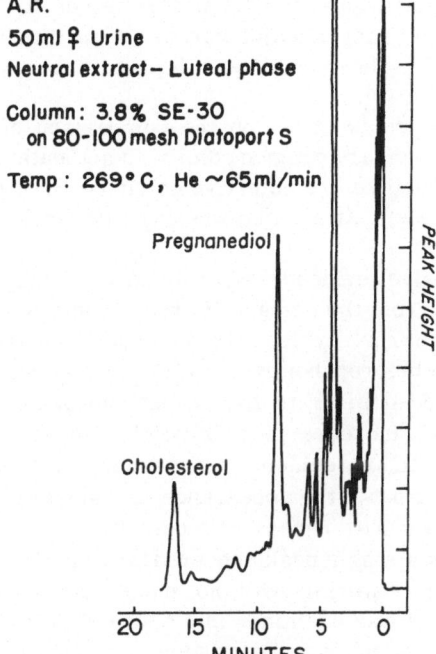

Figure 14-5. Gas chromatogram of a neutral luteal phase extract on an SE-30 column; $n = 4100$.

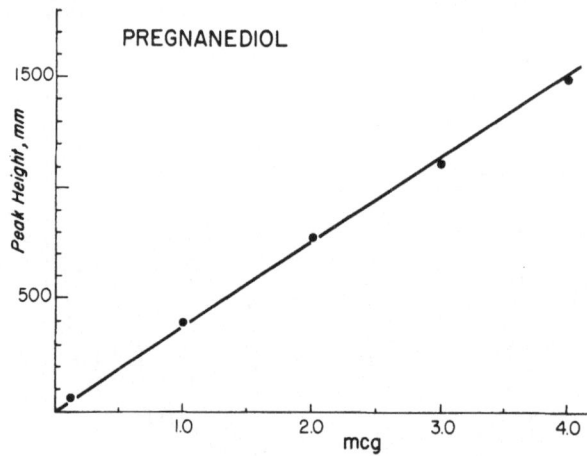

Figure 14-6. Response of flame detector to increasing amounts of pregnanediol diacetate. Each point represents the mean value of three injections.

but one (allopregnanediol) would be completely separated were they present in an extract. The latter is known to be present in urine, but in very low concentrations, and for most purposes would not contribute significantly to the pregnanediol values. This finding is in contrast to the report of Kirschner and Lipsett [4], who report values indicative of separation on SE-30.

Further evidence is based on the comparison of retention times for the presumed urinary pregnanediol with authentic material on both polar, nonpolar, and ketone retentive columns (Hi-Eff 8B, SE-30, and XE-60, respectively). At no time was any difference in retention time noted.

More conclusive evidence was obtained by trapping the proper effluent obtained from the extract of urine from a female in the luteal phase of the menstrual cycle. The material thus obtained was chromatographed as before; however, a small amount of pregnanediol diacetate was added prior to the second chromatogram to allow an estimated 30–50% increase in peak height. A splitting of this peak would present conclusive evidence for nonidentity of the two substances, while on the other hand, the appearance of a single symmetrical peak suggests, but does not necessarily constitute, proof of identity. On all three columns a single peak appeared following the addition of such authentic carrier material with no measurable peak spread. The remaining material was examined in a double-beam infrared spectrophotometer using a micro KBr window. A spectrum identical with that of authentic pregnanediol was obtained. The spectrum also showed a small number of extraneous peaks which, it was possible to show quite clearly, came from bleeding of the stationary phase (SE-30) (Figure 14-7). It is estimated that presence of impurity would have been detected to a level of approximately 10%.

The sensitivity of the method appears to be sufficient to allow the determination of $50\mu g$ per 24 hr of urine collection. The accuracy of the method was tested by the addition of pure pregnanediol, in concentrations varying from 50 to $5000\mu g$, to 1 liter aliquots of an unhydrolyzed urine pool (male) and analysis of each before and after addition of steroids. It can be seen from Table 14-7 that in all instances better than 90% of the added steroid was recovered. Table 14-8 shows multiple analyses of duplicate and quadruplicate aliquots of a number of different urines as an indication of the precision of the method.   In each instance the concentration of steroid obtained is well within the commonly accepted limits of precision for biochemical procedures.

Further important evidence for the validity of the method is obtained by comparison of excretion values obtained using GLC methods with those obtained by the Klopper procedure. Two publications indicate that at least in the range of about 1 mg per 24 hr and higher, there is excellent coincidence between the two methods (Tables 14-3 and 14-4). When pregnanediol excretion drops below the 1 mg per 24 hr value, the classical assay tends to give readings considerably greater than those given by the GLC method. It is likely, therefore, that at the very low levels the Klopper assay is less accurate due to the nonspecificity of the sulfuric acid chromogen. Typical chromatograms obtained from pregnancy and proliferative phase urines are shown in Figures 14-4 and 14-3.

TABLE 14-6. Separation of Some $C_{21}O_2$ Steroids on 3% SE-30*

Steroid	t_r (min)
3β -Hydroxy-5β -pregnan-20-one	7.0
5β -Pregnane-3α, 20α -diol	8.6
5α -Pregnane-3α, 20α -diol	8.7
Pregn-5-ene-3β, 20β -diol	9.2
5β -Pregnane-3β, 20α -diol	10.6
5α -Pregnane-3β, 20α -diol	10.8
5α -Pregnane-3β, 20β -diol	11.6
Pregn-5-ene-3β, 20α -diol	13.2

*The column was operated at 251°C with 55 ml of helium flow.

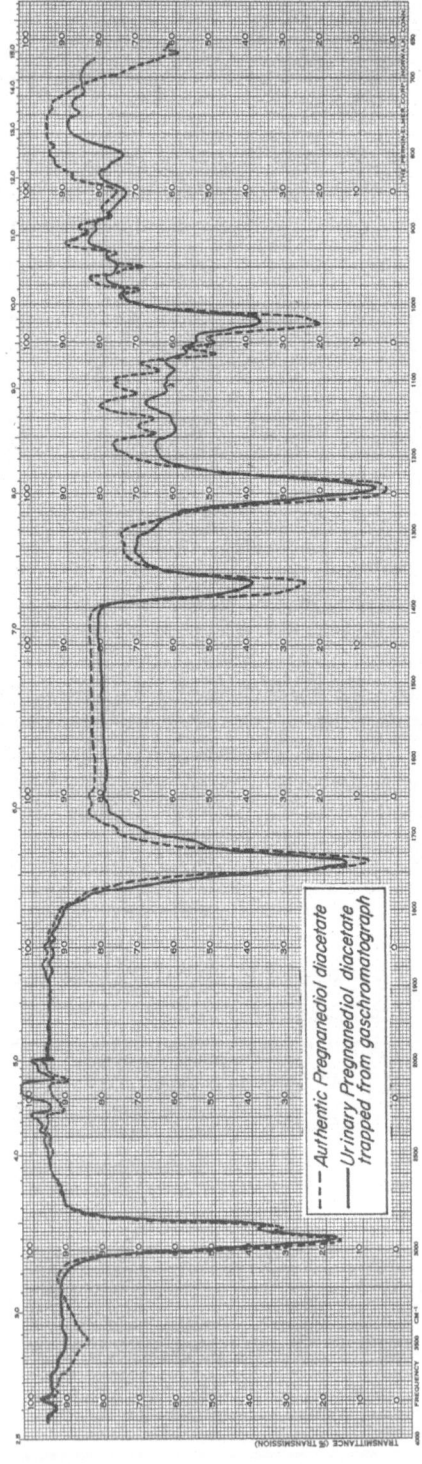

Figure 14-7. Comparison of infrared spectra of authentic pregnanediol diacetate and material trapped at the effluent stream of the gas chromatograph following injection of neutral luteal phase urine extract.

TABLE 14-7. Recovery of Pregnanediol Added to Urine *

Added (μg)	No. of experiments	Recovered, mean (μg)	Range of recovery (μg)	Standard deviation
50	6	48.6	46.7−50.2	±1.4 (±3%)
500	6	476	462−498	±13 (±2.7%)
5000	4	4920	4798−4980	±88 (±1.8%)

*Reproduced by permission from H. H. Wotiz and P. J. Mozden, Human Ovulation, edited by C. S. Keefer (Little, Brown and Company, Inc., Boston, 1965), Chapt. 10, Table 6. Sample: male urine pool.

TABLE 14-8. Analysis of Multiple Aliquots of Early Pregnancy Urines

Specimen	No. of determinations	Mean (mg)	Range (mg)	Standard deviation
A.H. 22-1962	4	5.7	5.3-5.9	±0.15 (±2.7%)
A.H. 25-1962	4	19.1	17.8-20.6	±1.2 (±6.3%)
A. H. 23-1962	4	7.7	7.1-8.2	±0.20 (±2.6%)
A.H. 20-1962	4	5.5	5.1-5.9	±0.21 (±3.8%)
E.B. 1-1965	2	3.85	3.83-3.86	
2-1965	2	4.18	4.15-4.20	
3-1965	2	4.85	4.81-4.89	
4-1965	2	8.50	8.45-8.55	
5-1965	2	10.8	10.7-10.9	
6-1965	2	16.4	16.1-16.6	
7-1965	2	28.2	28.0-28.4	
8-1965	2	14.1	13.8-14.3	
9-1965	2	11.4	11.3-11.5	

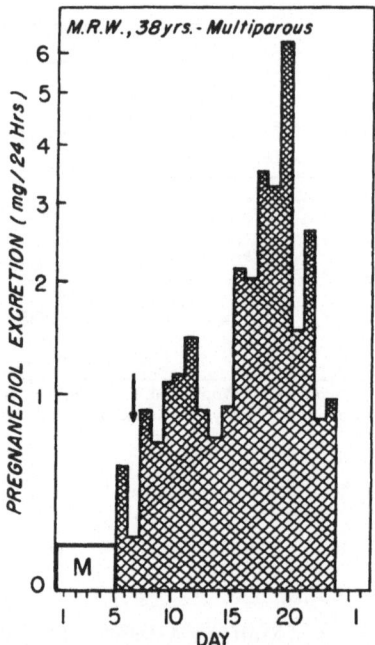

Figure 14-8. Pregnanediol excretion during a normal menstrual cycle as determined by gas chromatography. [H.H. Wotiz and P.J. Mozden, Human Ovulation, C.S. Keefer, ed. (Little, Brown and Co., 1965)].

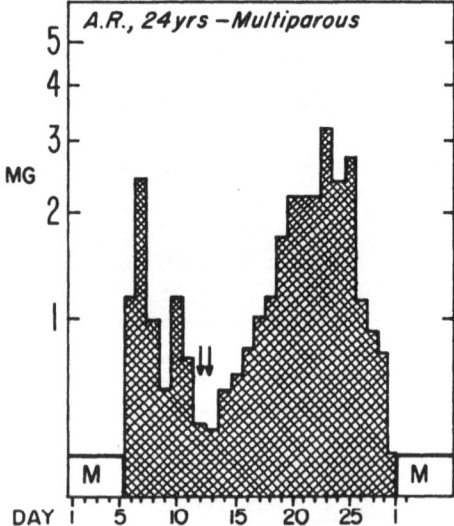

Figure 14-9. Pregnanediol excretion during a normal menstrual cycle as determined by gas chromatography. [H.H. Wotiz and P.J. Mozden, Human Ovulation, C.S. Keefer, ed. (Little, Brown and Co., 1965)].

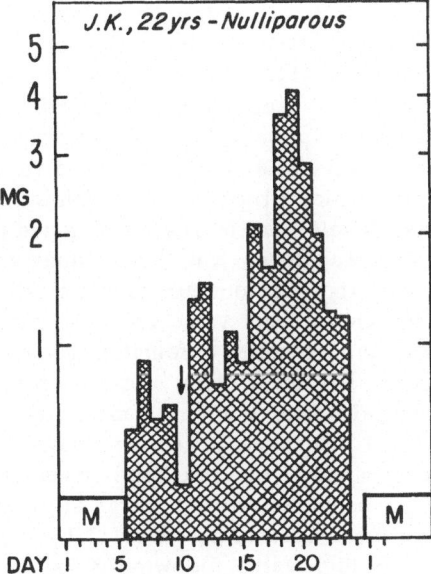

Figure 14-10. Pregnanediol excretion during a normal menstrual cycle as determined by gas chromatography. [H.H. Wotiz and P.J. Mozden, Human Ovulation, C.S. Keefer, ed. (Little, Brown and Co., 1965)].

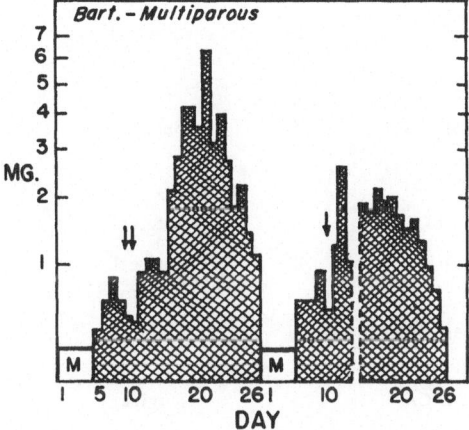

Figure 14-11. Pregnanediol excretion during two successive menstrual cycles as determined by GLC. The dotted lines indicate a lost specimen.

The application of the procedure described by us to a number of urines obtained during the normal menstrual cycle is shown in Figures 14-8 through 14-11. All the women were known to ovulate and were clinically devoid of any endocrine disorders. It is interesting to note that on a number of occasions somewhere in the early part of the cycle, but near mid-cycle, there is a 24 to 48 hr period in which excretion of pregnanediol drops to well below the accepted mean proliferative phase levels. In the results obtained for twelve normal menstrual cycles, it was found that these values varied between 150 and 200 μg and are thought to represent the adrenal base level of progesterone production. This value shows a good correlation with the figures obtained for a number of females treated with a sequential contraceptive drug. The mean value obtained for 29 women treated for a period of several weeks to several months was 178 μg per 24 hr.

Better data regarding adrenal base level have been obtained from 13 analyses of pregnanediol in urine from seven cancer patients six of whom had been oophorectomized surgically, and one by X ray. From Table 14-9 a mean adrenal base level of 161 μg per 24 hr can be computed. The lone high value following X-ray castration was well above this mean and may simply suggest that radiation castration often is inadequate.

Because of the well-known difficulties attendant to voluntary and accurate urine collection, it was thought useful to determine the

TABLE 14-9.   Pregnanediol Excretion After Bilateral Oophorectomy

Patient	Diagnosis	Age	Procedure	γ Pregnanediol per 24 hr
ADA	Breast CA	40	Surgical	201
				182
				148
ALP	Breast CA	65	Surgical	103
				200
				152
BAR	Breast CA	35	Surgical	193
MCC	Breast CA	47	Surgical	125
				223
TRA	Breast CA	55	Surgical	102
				129
ZAG	Breast CA	58	Surgical	172
			Mean	161 ± 40
OLEG	Breast CA	56	X-ray	450

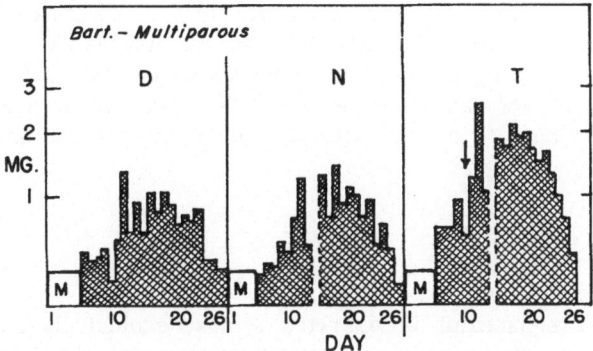

Figure 14-12. Comparison of diurnal and nocturnal pregnanediol excretion using GLC.
$D = 7$ a.m. – 7 p.m. collection; $N = 7$ p.m. – 7 a.m. collection; T = Summation of $D + N$.
Dotted lines indicate a lost specimen.

diurnal–nocturnal variation in pregnanediol excretion in the hope that
adequate data may be obtainable from overnight specimens. Figure
14-12 shows the results obtained when urine was collected in 12 hr
portions from 7 a.m. to 7 p.m. and 7 p.m. to 7 a.m. In general, when
the diurnal output was low the nocturnal collection showed increased
steroid concentrations and vice versa.

In 1957, Klopper [2] reported the fact that the administration of
ACTH caused an increase in the output of pregnanediol but that this
did not take effect in patients following adrenalectomy. In his book,
Clinical Application of Hormone Assay, Loraine [8] suggested that this
observation may make pregnanediol assays useful as a means of
determining adrenocortical function in man. Greater emphasis may be
placed on this point in view of the fact that this procedure may now be

TABLE 14-10. Effect of ACTH on Pregnanediol Excretion

Patient	Period	17-OH CS	17-KS	Pregnanediol
Apu	Control	4.2	6.0	0.54
	ACTH (1)	7.1	18.0	0.59
	ACTH (2)	20.1	22.7	0.77
Rot	Control	5.4	12.9	0.44
	ACTH	26.4	23.5	2.9
Bor	Control	6.5	9.4	0.38
	ACTH (1)	27.0	15.7	0.31
	ACTH (2)	22.3	27.7	1.2
Moore	Control	8.6	20.3	0.46
	ACTH (2)	26.3	46.8	0.82

carried out in somewhat less than 3 hr. Verification of the findings of
Klopper can be seen in Table 14-10. Four normal males were treated
with ACTH and pregnanediol excretion was measured before and after
administration of the hormone. A 1.5–6 fold rise in pregnanediol
levels was found to occur on the second day after ACTH administration.

Hydrolysis

Since pregnanediol is excreted almost exclusively as the glucosiduronate, it becomes necessary to liberate the lipid-soluble
steroid moiety. This can be done in one of two ways: hydrolysis by a
strong mineral acid, such as described by Klopper, or by the use of
the enzyme β-glucuronidase.

There is some evidence that a portion of the total pregnanediol is
destroyed by mineral acid treatment, while cleavage with enzyme
takes approximately 8–12 hr. Those analysts desiring rapid assays
must satisfy themselves with mineral acid treatment for 10 min.
On the other hand, anyone requiring the most accurate determination
will most likely utilize an enzyme preparation. Figure 14-13 shows
results obtained in our laboratory to determine the optimal concentration of enzyme needed. Maximal hydrolysis occurred at a concentration of enzyme of approximately 500 units per ml of urine using
the *helix pomatia* preparation "Glusulase" (Endo Products).

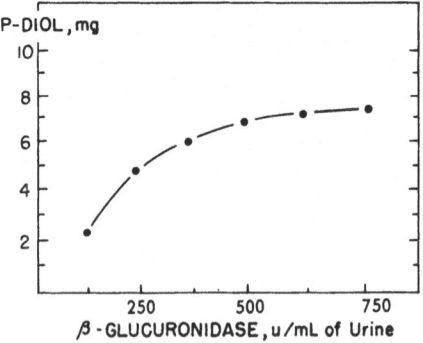

Figure 14-13. Enzymatic hydrolysis of urinary pregnanediol glucuronide. Two 50 ml aliquots
of a pregnancy urine from a pool were analyzed for the determination of each point. Analysis
was by GLC. β-Glucuronidase preparation used was "Glusulase" (Endo Products, Richmond
Hill, N.Y.).

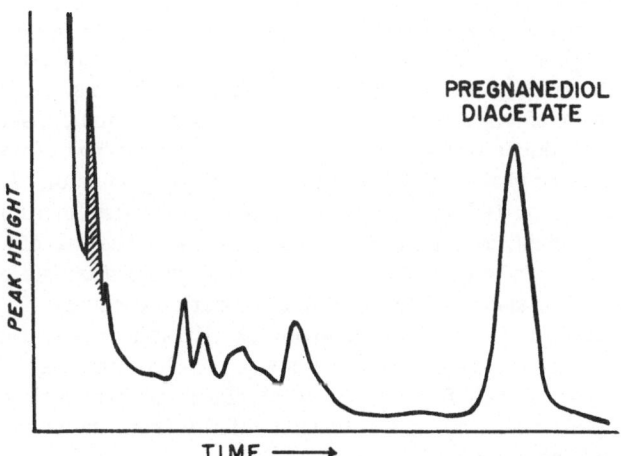

Figure 14-14. Gas chromatogram of pregnanediol diacetate from urine extract. GLC was performed on a portion of nonpregnancy urine after acetylation and thin-layer chromatography [Kirschner and Lipsett, Steroids 3:277 (1964)].

EXPERIMENTAL

Method P-I (Kirschner and Lipsett)

Hydrolysis and Extraction.   A $\frac{1}{10}$ aliquot of the urine collection is incubated with 500 units of glucuronidase (beef liver) per ml of urine for 24 hr at 37°C. The urine is transferred to a separatory funnel and extracted three times with 50 ml portions of ether. The combined ether extracts are then washed with 25 ml of 1 N sodium hydroxide, followed by two water washes of 25 ml each. The extract is dried over anhydrous sodium sulfate and after filtration the ether is evaporated to dryness on a rotating still. The dried crude extract is treated with 15 drops of acetic anhydride and 10 drops of dry pyridine. After heating to 56°C for 3 hr, the excess reagents are evaporated under a stream of dry air.   The residual material is applied as a narrow streak to a thin-layer plate.

Reference samples of pregnanediol diacetate are applied on the edges of the plate and the chromatogram is developed in the system benzene: ethyl acetate (80:20) for 45 min. Following air drying of the plate the surface is sprayed with an 0.1% solution of rhodamine 6G in ethanol.   The plate is then examined under ultraviolet light and the nonabsorbing areas (white spots on yellow background) corresponding to the standards are marked. The zone corresponding to pregnanediol is then scraped off the plate into a 15 ml centrifuge tube and the steroid eluted by triturating twice with 5 ml of acetone. The acetone extract is transferred to a small tube and evaporated to dryness with air or nitrogen. Following the addition of androstenedione as marker, the extract is ready for gas chromatography.

GLC of Pregnanediol. Gas chromatographic analysis is carried out on a 6-ft-long, 1% SE-30 column. The stationary phase is deposited on 60-80 mesh Gas Chrom P (Applied Science Laboratories). A flow rate of 25 ml/min of carrier gas is used. Kirschner and Lipsett do not give specific details of the temperature used, the method of deposition of the stationary phase, or the elution time of pregnanediol. A further complication reported in this paper is the very short linear range of the detector used, requiring trial runs to allow the use of a proper aliquot. In our experience this can be avoided by use of better instrumentation, specifically the flame ionization detector.

A typical chromatogram obtained by Kirschner and Lipsett is shown in Figure 14-14. Quantitation was carried out by establishing daily standard curves for pregnanediol diacetate and comparing the product of peak areas and the retention volume from the urine extract with those of the standard.

Method P-II (H.H. Wotiz)

Hydrolysis and Extraction. Fifty ml of urine is diluted to 150 ml and added to a round bottom flask containing 50 ml of toluene. A glass bead is introduced and the mixture is brought to a boil. At this point, 10 vol. % (15 ml) of concentrated HCl is added through the reflux condenser. The mixture is refluxed for 10 min and the flask immediately cooled with water. After transfer to a separatory funnel the toluene layer is removed and the urine is extracted once more with 50 ml of toluene. The organic layers are combined and washed successively with 2 × 25 ml of a 25% NaCl solution in 1 *N* NaOH and 2 × 25 ml of water. The toluene layer is transferred to a round bottom flask and reduced to dryness on a rotating still *in vacuo* at 60°C. The residue is transferred with acetone to a 1-dram screw-cap vial and again carefully taken to dryness in a stream of air or nitrogen. Then 0.1 ml of pyridine and 0.5 ml of acetic anhydride are added; the cap, with an aluminum foil insert, is tightened and the bottle is put in a water bath at 50°C. After 1 hr the mixture of solvents is evaporated at 50°C under a stream of nitrogen. Next, 100 μl of a solution of 10 mg of cholesterol (as the acetate) in 10 ml of acetone is added. For pregnancy urine 500 μl should be used. Care must be taken to ensure solution of the steroid from the wall.*

---

*It has been observed recently in another laboratory that some urine samples contain significant quantities of a substance having a retention time identical to that of cholesterol acetate. It is not yet known whether this material is present in the urine or whether it is an impurity in the solvents. In order to prevent errors arising from the presence of this material, quantitative measurement must be made by comparison with an external standard of pregnanediol.

Gas Chromatography

Column. . . . . .6 ft, stainless steel
($\frac{1}{8}$ in. OD) or glass
(2 mm ID) 3% SE-30 on
80-100 mesh diatoma-
ceous earth (acid wash-
ed, silanized)
Temperature. . . . . .250°C
Flow . . . . .70 ml/min
Flame ionization detector temperature. . .300°C
Glass lined injector . . . . .280°C or,
On-column injector. . . . . .250°C or,
Micropipette injector . . . . .280°C

Priming. Before an analysis should be attempted, the column must be saturated with pregnanediol and cholesterol acetates each day. Three injections of about 10 μg of standard in rapid order generally suffice. Thereafter, excellent reproducibility (± 1%) may be attained.

Injection. Usually 2 μl is drawn into the capillary pipette or a Hamilton syringe. In the case of the former, a small tissue (Wipette) is wetted with solvent and the outside of the pipette is wiped gently. Care must be taken not to touch the orifice or the sample will discharge. The amount to be injected depends largely on the concentration of the components to be analyzed. Other factors must, however, be taken into consideration. An excessive injection may cause overlapping of peaks and may obscure minor peaks. It may also prevent the detector from approaching baseline to give readings at a sufficiently low attenuation. Generally, it is more desirable to inject small volumes and decrease the attenuation.

Measurement. The ratio of peak heights for pregnanediol diacetate and cholesterol acetate must be determined daily since even with the same detector and column some variation can be found.

$$r_p = \frac{\text{peak height of cholesterol acetate}}{\text{peak height of pregnanediol diacetate}}$$

The concentration of pregnanediol is determined by the formula

$$C_u = \frac{H_u \times 100 \times r_p}{H_{chol}} \times \text{aliquots}$$

where C_u is the concentration of pregnanediol in the urine, H_u is the peak height of pregnanediol in the urine extract, and H_{chol} is the peak height of cholesterol acetate.

DETERMINATION OF PROGESTERONE

Evaluation of Methods

The chemical determination of progesterone in blood or blood plasma probably ranks among the most difficult of assay techniques. The method of Zander [9], which allows the determination of 5 μg of progesterone per 100 ml of blood, showed that the normal progesterone levels in humans are well below the limits of sensitivity of this procedure. Zander was able to measure the concentrations of progesterone during the second half of pregnancy as well as in placental tissues, corpora lutea, and follicular fluid. Similar sensitivities were found for other assay procedures developed by Short [10] and Oertel et al. [11]. The application of gas chromatographic separation, following some form of preliminary purification such as thin-layer chromatography, has opened the way for considerable improvements in both the speed and the sensitivity of these procedures.

As early as 1962, Kumar et al. [12], using gas chromatographic separation and identification techniques, were able to demonstrate the presence of progesterone in pregnancy myometrium. Futterweit et al. [13] have been successful in determining the plasma concentrations of progesterone beginning at the fourth month of pregnancy; their values appear in rather good agreement with those published by Oertel.

More recently, Nishizawa and Eik-Nes [14] characterized progesterone and androstenedione obtained from the dog ovarian artery following the administration of chorionic gonadotropin.

In 1964, Yannone et al. [15] published a general method for the determination of progesterone in human plasma. As a result of the use of solid state injection system, concentrations of 0.024 μg of progesterone per sample could be measured with a precision of 10%. Preliminary purification is achieved through paper chromatography after dilution with radioactive progesterone used to locate the steroid on the paper and to determine the amount of recovery. Gas chromatography is carried out on a rather short XE-60 (ketone retentive) column, probably to attain rapid elution to allow sufficient sensitivity.

The overall recovery is reported to be 34 ±14% in 30 experiments. Linearity for the flame detector was amply demonstrated between 0.003 and 1.00 μg of progesterone. Specificity of the procedure is based exclusively on the paper chromatogram and gas chromatography on XE-60, QF-1, and SE-30 silicone polymers. The precision of the method appears to be quite good as seen in Table 14-11. A typical set of chromatograms obtained by the authors is shown in Figure 14-15.

Collins and Sommerville [16] very recently reported a much more elaborate procedure utilizing thin-layer chromatographic purification

TABLE 14-11.   Analysis of Duplicate Samples from Fifteen Patients

Sample size	Source of sample	Measured concentration in μg per 100 ml		% Deviation from the mean
		Sample I	Sample II	
1	39 Weeks gestation	13.8	13.4	1.47
3	38 Weeks gestation	12.9	14.8	6.86
3	37 Weeks gestation	16.0	17.0	3.03
3	36 Weeks gestation	20.5	20.1	0.99
5	33 Weeks gestation	13.0	15.0	7.14
4	29 Weeks gestation	10.0	9.3	3.63
6	29 Weeks gestation	10.0	12.0	9.09
4	27 Weeks gestation	7.2	5.4	14.29
4	24 Weeks gestation	7.8	7.8	0.00
4	20 Weeks gestation	4.7	5.4	6.93
4	20 Weeks gestation	6.8	8.3	9.93
5	17 Weeks gestation	4.6	4.5	1.10
15	8 Weeks gestation	3.4	3.0	6.25
15	Luteal phase Menstrual cycle	2.3	2.7	8.00
20	Anovulatory hirsute Female	0.26	0.19	15.56

*Reproduced by permission from M. E. Yannone, D. B. McComas, and A. Goldfien, J. Gas Chromatog. 2:30 (1964), Table II.

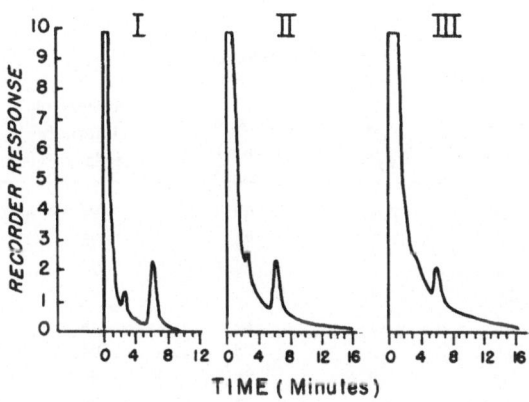

Figure 14-15. Gas chromatograms obtained in the analysis of: (I) 5 ml of plasma from a 34 week gestation. Attenuation 300. Peak area is equal to a progesterone mass of 0.25 μg. (II) 10 ml of plasma from a 10 week pregnancy. Attenuation 100. Peak area is equal to progesterone mass of 0.12 μg. (III) 5 ml of plasma from the luteal phase of the menstrual cycle. Attenuation 30. Peak area is equal to a progesterone mass of 0.002 μg. The retention time of progesterone is 6 min. [Yannone, McComas, and Goldfien, J. Gas Chromatog. 2:30 (1964)].

followed by gas-radiochromatographic quantification. They claim to be able to complete an analysis in approximately 4 hr with sufficient sensitivity to allow the determination of progesterone in 10 ml plasma samples during the menstrual cycle.

It should be pointed out, however, that this particular procedure requires equipment in addition to a standard gas chromatograph to allow the simultaneous mass measurement and determination of radio-activity. This is achieved by combustion of the effluent vapor which is then passed through a proportional counter coupled to a rate meter. The advantage of such a procedure is, of course, the ability to determine the total recovery of progesterone for each individual analysis. Solid state injection by a method similar to that described by Menini [17] allows greater sensitivity of mass measurement. The drawback of the method, as it stands reported presently, is the fact that the radio-active progesterone marker is added in amounts larger than the total

TABLE 14-12.   Female Plasma*

(a) Early pregancy
(replicates on pool)

Recovery of internal standard (%)	Progesterone (μg per 5 ml aliquot)
84	0.22
100	0.21
20	0.22
85	0.19

(b) Menstrual cycle

Subject No.	Age (years)	Day of cycle	Duration (days)	Recovery of internal standard (%)	Progesterone (μg per 100 ml)
1	28	4	32	24	0.52
2	23	9	29	95	0.75
3	24	10	30	76	0.97
4	18	13	32	96	0.9
5	17	17	29	12	3.9
6	34	19	28	101	1.4
7	35	19	28	90	2.3
8	38	21	27	99	0.6
8	38	23	27	100	0.53
4	18	24	30	100	1.1
3	24	26	30	94	1.1
9	25	28	35	80	1.1

*Reproduced by permission from W. P. Collins and I. F. Sommerville, Nature 203:836 (1964).

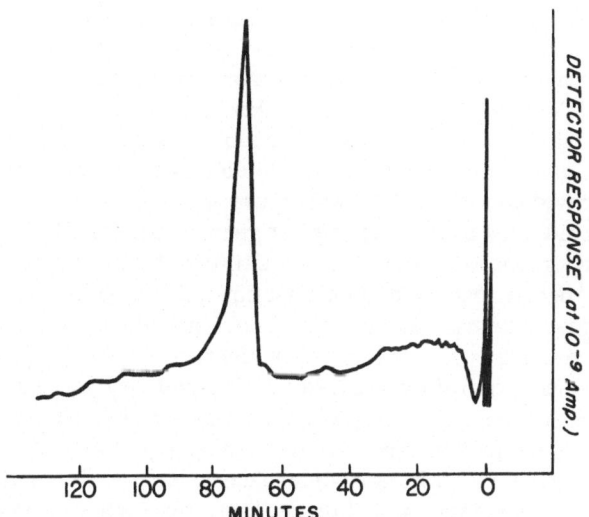

Figure 14-16. GLC of progesterone (endogenous and internal standard) eluted from TLC plate on 1% CDMS at 210°C and using solid injection [W. P. Collins and I. F. Sommerville, Nature 203:836, (1964)].

amount of material present in the plasma sample. It is expected, however, that this difficulty can be overcome by the utilization of progesterone-4-^{14}C with much greater specific activity.

Accuracy, based on recovery of labeled internal standard, varied from 12% to 101%. Despite the addition of such large amounts of progesterone-^{14}C and the great variability in recovery, the precision of the method appears to be good.

The concentrations of progesterone found in a number of female subjects at different times of the menstrual cycle are shown in Table 14-12. Specificity of the procedure is based on two-dimensional thin-layer chromatograms and on gas chromatography on the CDMS column as well as on neopentylglycol adipate, QF-1, and SE-30. Further evidence of specificity was obtained by comparison of the retention times for the dimethylhydrazone and bis-ethylene ketal of the extracted progesterone with the derivatives of the authentic material. A typical chromatogram obtained by these investigators is shown in Figure 14-16.

Because of the rather expensive equipment necessary for this procedure and the inherent difficulty pointed out above, as well as the availability of two methods not requiring this type of instrumentation, a detailed description will not be presented.

A very elegant procedure for the measurement of plasma progesterone using the electron capture detector has recently been described

by VanderMolen and Groen [18]. The ability of certain organic compounds to capture thermal electrons varies greatly depending on the
amount of unsaturation, nature of the ring structure, and substitution
with certain elements (halogens).   Except for highly unsaturated
molecules having at least three double bonds, steroids may be
considered as substances showing little electron attachment. The few
methods previously described which use electron capture measurement of steroids, all utilize derivatives generally formed by interaction
of a hydroxyl group with a halogen-containing alkyl radical.

Since progesterone is a monounsaturated diketone, it has a very
low electron-capturing coefficient and does not allow the formation of
volatile derivatives with the desired electron-capturing properties.
The authors have therefore utilized the 20β-hydroxy steroid dehydrogenase from *Streptomyces hydrogenens* to reduce the carbonyl group at
C-20, forming exclusively the 20β-hydroxy-pregn-4-ene-3-one. This,
in turn, is reacted with monochloroacetic anhydride as described
originally by Landowne and Lipsky [19], producing a moderately
strong electron-capturing derivative:

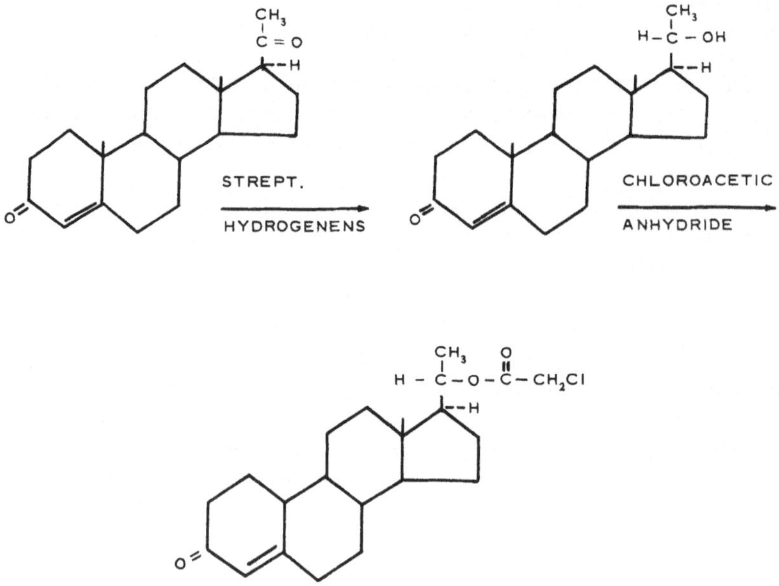

Since the measurement of blood progesterone involves the analysis
of extremely minute samples and recovery becomes somewhat
uncertain from sample to sample, the authors begin the assay procedure

by the addition of a known concentration of 7-^{3}H-progesterone before ether extraction of the alkalinized plasma. After thin-layer separation of the progesterone, the eluate is reduced with the dehydrogenase and chloroacetylation follows. Further thin-layer chromatography of the chloroacetate obtains sufficient purity to allow its gas chromatographic separation and detection.

The utilization of the electron capture detector requires somewhat greater caution because of its enhanced sensitivity as well as its smaller range of linearity. The authors were able to demonstrate a linear range between 0.001 and 0.200 μg.

In order to ascertain losses on the column or by transfer, testosterone chloroacetate was used as the reference standard throughout the technique. In a series of 29 experiments over a period of several months, the peak ratio of the reference standard to the progesterone derivative remained at 1.15 with a standard deviation of 4%.

Accuracy of the procedure was studied by the addition of tritiated progesterone to water and to plasma, and carrying this mixture through the complete assay procedure. It is important to note that extensive incubation of progesterone with the 20β-dehydrogenase results in a considerable loss of C-20 reduction product and exposure to the enzyme has therefore been limited to 2 hr. In 22 experiments, 48.9% of the added steroid was recovered with a standard deviation of 5.5% and a range of 44.6–57.8%. When the enzyme incubation was extended to 16 hr, only 16.7% of the added steroid was recoverable. The authors estimate that if this technique is used, approximately 0.02 μg per 100 ml of plasma can be detected; in fact, they have been able to measure several plasma samples at that level.

With respect to precision it appears that at concentrations of 0.1 μg per 100 ml of plasma or less, a variance of up to 40% may be encountered. With larger samples a variance of from 10–15% was reported, and in specimens containing 1 μg per 100 ml of plasma or greater, a variance of approximately 5% was attained. Improvement in the precision may of course be obtained by use of multiple injections from the same sample and averaging the results.

As must be expected from new methods, particularly when such minute amounts of substance are measured, proof of the specificity of the reaction is somewhat equivocal. Nevertheless, the authors have presented evidence that is highly suggestive of the specificity of the method described. This evidence is based in part on the thin-layer chromatographic separation of progesterone from a number of other steroids which might conceivably interfere with the analysis (Table 14-13). Further strong evidence for the specificity is based on the enzymatic reduction, chloroacetylation, and thin-layer and gas chromatography of the acetylated product (Table 14-13). The fact that on a ketone retentive column (XE-60) identical retention times for the extracted material and authentic standard have been obtained is significant. Endocrinologic evidence for the specificity was obtained by examination of several plasma samples from ovariectomized, adrenalectomized women, none of whom showed the presence of 20β-hydroxy-pregn-4-en-3-one chloracetate. The values obtained by these authors on nine normal male subjects varied from less than 0.02 to $0.049\mu g$ per 100 ml. During the proliferative phase, values of less than 0.01 to 0.079 μg per 100 ml of plasma were found, while during the luteal phase, between 0.60 and 1.99 μg was measured. A typical chromatogram obtained is shown in Figure 14-17.

TABLE 14-13. Thin-Layer Chromatography and Gas Chromatography of Steroids *

Steroid	Thin-layer chromatography		Gas chromatography on 1% XE-60 Relative† retention of chloroacetates
	R_f in benzene: ethyl acetate (2:1)	R_f after chloroacetylation in benzene: ethyl acetate (6:1)	
Pregnanediol	0.18	0.91	39.0
Testosterone	0.26	0.42	21.4
20α-Hydroxypregn-4-en-3-one	0.32	0.47	33.2
20β-Hydroxypregn-4-en-3-one	0.32	0.52	29.7
17α-Hydroxyprogesterone	0.33	0.12	—
Etiocholanolone	0.35	0.56	12.0
Androsterone	0.40	0.55	9.3
3β-Hydroxypregn-5-en-20-one	0.43	0.70	15.2
Dehydroepiandrosterone	0.44	0.59	10.9
Androst-4-ene-3,17-dione	0.45	0.18	—
3α-Hydroxy-5β-pregnan-20-one	0.46	0.71	13.0
Estradiol	0.57	0.82	45.6
Progesterone	0.59	0.25	—
Estrone	0.88	0.56	17.5

*Reproduced by permission from H. J. VanderMolen and D. Groen, Gas Chromatography of Steroids in Biological Fluids, edited by M. B. Lipsett (Plenum Press, New York, 1965), p. 162, Table 3.
†Retention relative to that of 5α-cholestane = 1.

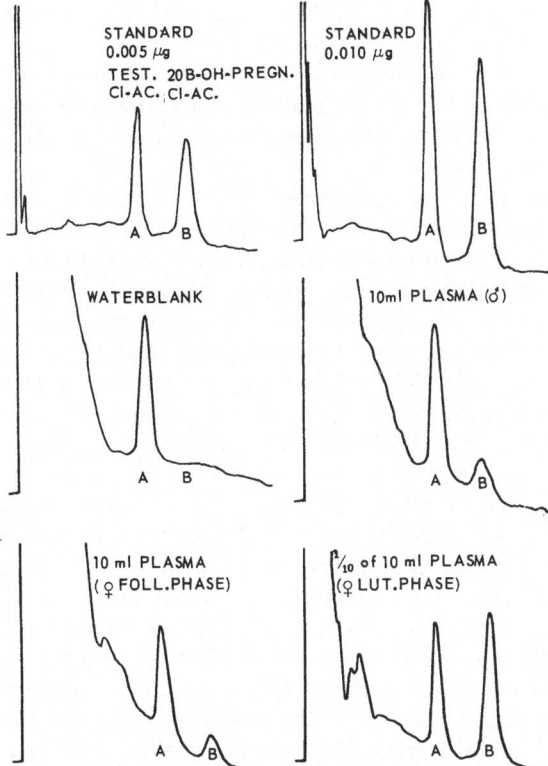

Figure 14-17. GLC of standards and samples taken through the method for the estimation of plasma progesterone. A = testerone-chloroacetate peak; B = 20β-hydroxy-4-pregnene-3-one-chloroacetate peak.

EXPERIMENTAL

Method PG-I (Yannone, McComas, and Goldfien)

Blood is drawn into a heparinized tube and immediately centrifuged. (After separation of the plasma it may be stored at -15°C.) To the plasma is added 10 μl of a progesterone-7-^{3}H ($\sim$ 25 mC/mg) solution containing approximately 0.001 μg per 10 μl of 95% ethanol ($\sim$ 15,000 cpm). Equal amounts of progesterone-7-^{3}H are pipetted to two counting vials to determine the exact amount of radioactive material added. The plasma lipids are then extracted by shaking gently with 3 volumes of ether for 10 min. After centrifugation and transfer of the ether layer, the solvent is evaporated at 40°C under air flow. To the residue is added 5 ml of warm 70% methanol. This solution then is stored in a freezer (−15°C) overnight to allow precipitation of lipids. The latter are separated by centrifugation at −15°C for 30 min at 2000−3000 rpm, followed by transfer of the supernatant to a clean test tube. An equal volume of heptane is added and the tube is shaken. After separation, the heptane is twice washed with $\frac{1}{2}$ volume of 70% methanol and the combined methanolic residues are reduced in volume to 30% by evaporation in a water bath under air flow. The aqueous residue is shaken with 5 ml of light petroleum ether and after removal of the organic phase the latter is reduced in volume to prepare for paper chromatography.

The material is now transferred to a 5-mm paper strip (Whatman No. 1, methanol washed). After the paper is hung in a chromatography jar containing 70% methanol—heptane, it is left to equilibrate at 23°C for at least 2 hr. After this period, heptane is introduced and the chromatogram is developed for 2 hr. After drying in air, the strip is scanned on a radioactivity chromatogram scanner and the area of labeled steroid thus detected is cut out within narrow limits and eluted with 30 drops of an equal mixture of benzene—methanol. The eluate is directly entered into a 21-mm-diameter glass tube 60 mm long with the last 40 mm tapered to a point. The solvent is evaporated as before, and 0.10 ml of 95% ethanol is added, care being taken to wash the material down the sides. In order to determine recovery of the carrier material, $10\,\mu l$ of this solution is transferred to a counting vial while the remaining solution is once again evaporated to dryness.

A relatively tedious transfer of the residue to a solid state injector by four successive washes with $10\,\mu l$ of solvent each is then recommended. After sample introduction the injector is again washed with solvent as is the vial and these washings are placed into another counting tube. (We suggest consideration of the alternate procedure of solid state sample introduction shown in Figure 4-9.)

In either case, the sample thus prepared is ready for gas chromatography. A 28-in.-long stainless steel column packed with 1% XE-60 applied on Gas Chrom CLH (60-70 mesh) is cured overnight at 230°C. Carrier gas at a pressure of 24 psi with the column at 195°C is used for chromatography, while the vaporizer is kept at 200°C.

Serial concentrations of progesterone are determined daily to allow quantification. Measurement of peak area (peak height × width at $\frac{1}{2}$ height) is recommended.

Fifteen milliliters of phosphor solution [4 g of 2,5-diphenyloxazole and 40 mg of 1,4-bis-[2-(5-phenyloxazoly)]-benzene in 1000 ml of distilled toluene] are now added to the vials set aside for counting. The samples are counted in a liquid scintillation spectrometer.

The concentration of progesterone can then be determined by the following formula:

$$\frac{S}{(A \times 9) - V} \times \frac{C}{P} \times 100 = \mu g \text{ progesterone per 100 ml of plasma}$$

where S is cpm of progesterone-7-^{3}H added to plasma, A is cpm of

progesterone-7-^{3}H in 0.1 aliquot of final extract, V is cpm of pro-
gesterone-7-^{3}H from syringe and vial after injection, C is μ g of
progesterone determined by GLC, and P is volume of plasma analyzed.

Method PG-II (VanderMolen and Groen)

Materials. Ether. Reagent grade, distilled through a 60-cm fractionating column.

Methanol. Anhydrous anlytical reagent, distilled from 2,4-dinitro-
phenylhydrazine.

Benzene. Thiophene free, analytical grade reagent. This solvent
is repeatedly washed with concentrated sulfuric acid followed by
water. It is then distilled twice through a 100-cm Vigreux column.

Ethyl Acetate. Analytical grade, distilled through a 60-cm
fractionating column.

Tetrahydrofuran. Refluxed for 3 hr with potassium hydroxide
pellets and distilled. It is then distilled from sodium and stored over
sodium in a dark brown bottle.

Pyridine. Refluxed over barium oxide from 4-6 hr and distilled
through a fractionating column. It is stored over calcium chloride in
a desiccator.

Chloroacetic Anhydride. Dried and stored in a desiccator.

Scintillation Fluid. 2,5-Diphenyloxazole (PPO) and 1,4-bis-[2-(5-
phenyloxazolyl)]-benzene (POPOP), scintillation grade. Used without
further purification.

Silica Gel TF 254. (Merck, according to Stahl, containing ultraviolet fluorescence indicator.) Used without additional purification.

Phosphate Buffer. 0.15 M of phosphate buffer at pH 5.2 was prepared by mixing 97.5 ml of 0.15 M potassium dihydrogen phosphate with
2.5 ml of 0.15 M disodium hydrogen phosphate solution. To 100 ml of
this buffer 100 mg of EDTA is added.

Tris-(hydroxymethyl—amino methane) Buffers.

0.1 M pH 8.1 without EDTA.   50 ml of an 0.1 M solution of
tris-hydroxymethyl—amino methane and 26 ml 0.1 N HCl are mixed and
diluted with distilled water to make 100 ml total volume.

0.005 M at pH 8.2 with EDTA. A 0.005 M buffer solution is
prepared by dissolving 300 mg of Tris in 500 ml of distilled water.
The pH is adjusted to 8.2, using a meter, by addition of 0.01 N HCl. To
100 ml of this buffer solution is added 100 mg of EDTA to make a
final concentration of 2.7×10^{-3} M.

Cofactor Solution. 5 mg of DPNH is dissolved in 3 ml of 0.1M
tris buffer at pH 8.1 without EDTA. (The solution can be stored at
4°C for one week.)

Enzyme Solution. The concentrated suspension of 20β-hydroxy steroid dehydrogenase prepared from *Streptomyces hydrogenens* (Boeh-ringer, Mannheim, Germany) is stable for several months if stored at 4°C. For use, a volume corresponding to approximately 4 mg of protein per ml of the concentrated suspension is diluted with 4 volumes of 0.005 M tris buffer at pH 8.2 containing EDTA.

7-^3H-Progesterone. Tritiated progesterone with a specific activity of approximately 10 C/mM was used. Prior to the use of tritiated progesterone, thin-layer chromatography in the systems benzene: ethyl acetate (1:1), cyclohexane: ethyl acetate (1:1), and benzene: ethyl acetate (4:1) is used to test the purity of the material.

Testosterone Chloroacetate. Prepared as follows: A solution containing 1 g of chloroacetic anhydride, 0.2 ml of pyridine, and 100 mg of testosterone in 5 ml of tetrahydrofuran is left standing for 18 hr in the dark. Following this, 5 ml of water is added and the derivative extracted 3 times with 5 ml of ethyl ether. The combined ether extracts are washed with 5 ml of distilled water. After the ether extract is filtered through anhydrous sodium sulfate it is evaporated to dryness. Following several recrystallizations of the amorphous residue, crystals with a melting point of 124° — 125°C (corrected) may be obtained.

Thin-Layer Chromatography. Silica gel plates (20 × 20 cm) are prepared to a thickness of approximately 0.30 mm. A mixture of 3 g silica gel in 9 ml of water is used. Plates are activated at 100° for 90 min prior to use. Samples are applied from chloroform: methanol (1:1) solution in 2 cm streaks, separated from each other by 1 cm. Following ascending development the standards are detected by means of a Haynes fluorescence scanner and the areas corresponding to the appropriate standard are scraped from the plate with a spatula. C a u t i o n: Since only trace amounts of steroids are being analyzed, clean tanks and fresh solvent mixtures must be used for each chromatographic plate.

Gas-Liquid Chromatography. Columns approximately 3 ft long and 0.4 cm ID are filled with 80–100 mesh Gas Chrom P coated with 1% XE-60 polymer. Chromatography is carried out at between 215° and 220°C with the flash heater operated at 250°C. The flow rate of carrier gas (nitrogen or helium) should be set to approximately 75 ml/min. Argon with 10% methane at a flow rate of 225 ml/min is used as a purge gas for this specific detector. Injections are made using a 10 μl Hamilton syringe. Quantification is carried out by determination of peak areas.

Measurement of Radioactivity. The samples to be analyzed are evaporated in glass vials and the residues dissolved in 0.1 ml of ethanol to which 10 ml of scintillation fluid is added. The samples are then counted in a liquid scintillation spectrometer for a sufficiently long time to allow a precision of between 1 and 2%.

Glassware. Because of the extremely low concentrations of steroids to be analyzed, particular care must be taken in the cleaning and preparation of glassware to prevent elution of unrelated material and adsorption of progesterone on the glass surface.

Glassware is soaked overnight in chromic acid and rinsed with tap water followed by soaking in detergent, rinsing again with tap water, followed by soaking overnight in dilute HCl, rinsing 10 times with tap water and 10 times with deionized water. The tubes are finally dried at room temperature. The 2 ml tubes used for collection of samples for GLC are siliconized with a 5% solution of dimethyldichlorosilane in benzene.

Method of Assay. Ten milliliters of plasma is added to a 50 ml glass tube containing approximately 10,000 counts/min of 7-^{3}H-progesterone. After the plasma is treated with 0.25 ml of 20% sodium hydroxide solution, the mixture is extracted 6 times with 15 ml of ether. The pooled ether extracts are washed twice with 5 ml of water, evaporated to dryness, and the residue is chromatographed on a silica gel thin-layer plate in benzene: ethyl acetate (2:1). The area corresponding to authentic progesterone is scraped off and extracted with 95% methanol (3 × 1 ml). After evaporation of the methanol and its concentration in the tip of a 15 ml centrifuge tube, the residual material is dissolved in 1 drop of ethanol, and 0.5 ml of the 0.1 M phosphate buffer at pH 5.2, as well as 0.03 ml of the cofactor solution and 0.03 ml of the diluted enzyme solution, is added. The contents are thoroughly mixed and incubated for 2 hr at 37°C. Following this, 1 ml of distilled water is added and the solution is extracted with 4 × 1 ml of ethyl acetate. The combined extract is again evaporated under nitrogen, and the tubes put in a vacuum desiccator for about 3 hr. To this residue is then added 0.5 ml of a solution of monochloroacetic anhydride in tetrahydrofuran (10 mg per 10 ml) and 0.1 ml of pyridine, and the tube is replaced in the desiccator overnight. The reaction is stopped with 1 ml of distilled water and the steroid chloroacetate extracted three times with 1 ml of ethyl acetate. The combined extracts are washed once with 1 ml of 6 N HCl and twice with 1 ml of distilled water, following which the extract is evaporated to dryness. After application of the residue to a silica gel thin-layer plate and development in a benzene: ethyl acetate system (6:1), the area corresponding to the standard of 20β-hydroxy-pregn-4-en-3-one chloroacetate is scraped off and 1 ml of benzene is added to the silica gel. After the extract is mixed thoroughly, 0.02 ml of distilled water is added, agitation is repeated, and the immiscible layer is cleared by centrifugation.   The benzene layer is transferred to a 2 ml siliconized conical centrifuge tube and dried down after each of three extractions. This residue is then dissolved in 1 ml of methanolic testosterone chloroacetate and the contents of the tube thoroughly mixed. (During the luteal phase of the menstrual cycle 0.04 μg of testosterone chloroacetate is added; in instances where less progesterone would be expected, 0.01 μ g of testosterone is added.) One-tenth of the diluted material is now removed for radioactive counting and the remainder again taken to dryness, taking care to concentrate the material in the tip of the tube. The residue thus obtained is dissolved in 10 or 20 μl of benzene (depending on the amount of internal standard added) and 5 μl of the extract is injected into the gas chromatograph.

The amount of progesterone in the plasma samples can then be calculated by the equation

$$P = R \times C \times U \times A \times 0.80$$

where

$$R = \frac{\text{cpm of 7-}^3\text{H-progesterone added to plasma}}{10 \times \text{cpm of tritium in aliquot prior to gas chromatography}}$$

$$C = \frac{\text{peak area of 0.01 } \mu g \text{ testosterone chloroacetate}}{\text{peak area of 0.01 } \mu g \text{ 20}\beta\text{-hydroxy-pregn-4-en-3-one}}$$

$$U = \frac{\text{peak area of 20}\beta\text{-hydroxy- pregn-4-en-3-one in sample}}{\text{peak area of testosterone chloroacetate in sample}}$$

$A = \mu$g testosterone chloroacetate added

$$0.80 = \frac{\text{molecular weight of progesterone}}{\text{molecular weight of 20}\beta\text{-hydroxy-pregn-4-en-3-one chloroacetate}} = \frac{314}{393}$$

Further correction must be made for the total mass of the tritiated progesterone added. This mass may be obtained by adding equal amounts of the progesterone to 10 ml of water and processing exactly as the plasma. The mass obtained in this manner is then subtracted from that calculated by the above equation.

REFERENCES

1. G. W. Corner and W. M. Allen, Am. J. Physiol. 88:326 (1929).
2. A. Klopper, J. A. Strong, and L. R. Cook, J. Endocrinol. 15:180 (1957).
3. A. Klopper, E. A. Michie, and J. B. Brown, J. Endocrinol. 12:209 (1955).
4. M. A. Kirschner and M. B. Lipsett, Steroids 3:277 (1964).
5. H. H. Wotiz and P. J. Mozden, Human Ovulation, edited by Keefer (Little, Brown and Co., Boston, 1965), p. 160.

6. H. H. Wotiz, Biochim. Biophys. Acta 69:415 (1963).
7. R. T. Cox, J. Chromatog. 12:242 (1963).
8. J. A. Loraine, Clinical Application of Hormone Assay (William & Wilkins, Baltimore, 1958), p. 235.
9. J. Zander and H. Simmer, Klin. Wochschr. 32:529 (1954).
10. R. V. Short, J. Endocrinol. 16:415 (1958).
11. G. W. Oertel, S. P. Weiss, and K. B. Eik-Nes, J. Clin. Endocrinol. Metab. 19:213 (1954).
12. D. Kumar, J. A. Goodno, and A. C. Barnes, Nature 195:1204 (1962).
13. W. Futterweit, N. L. McNiven, and R. I. Dorfman, Biochim. Biophys. Acta 71:474 (1963).
14. E. E. Nishizawa and K. B. Eik-Nes, Biochim. Biophys. Acta 86:610 (1964).
15. M. E. Yannone, D. B. McComas, and A. Goldfien, J. Gas Chromatog. 2:30 (1964).
16. W. P. Collins and I. F. Sommerville, Nature 203:836 (1964).
17. E. Menini and J. K. Norymberski, Biochem. J. 95:1 (1965).
18. H. J. VanderMolen and D. Groen, Airlie House Conference, Warrenton, Va., February 1965.
19. R. A. Landowne and S. R. Lipsky, Anal. Chem. 35:532 (1963).

Chapter 15

Determination of Androgens and Related Steroids

DETERMINATION OF 17-KETOSTEROIDS
Biochemistry

Androgens and 17-ketosteroids are derived from several organs. In the testes, the primary biosynthetic product appears to be the steroid testosterone.

TESTOSTERONE ⇌ Δ⁴-ANDROSTENE-3,17-DIONE

Recently, excellent evidence has been obtained that the ovaries of a variety of animals also participate in the synthesis of androgens. Its apparent primary $C_{19}O_2$ steroid is Δ^4-androstene-3,17-dione. The adrenal gland, while also involved in the biosynthesis of androstenedione, is more strongly associated with the synthesis of significant amounts of dehydroepiandrosterone as the sulfo-conjugate. The three above-mentioned products exhibit varying degrees of biological activity, testosterone being the most potent. Whether or not the biologic activity of the other two substances is due to their conversion to testosterone is presently unknown. Another C-19 steroid produced in the adrenal gland contains a further site of oxidation, at C-11, and is known as adrenosterone.

DEHYDROEPIANDROSTERONE   ADRENOSTERONE

In the course of metabolism of the $C_{19}O_2$ steroids, a series of reductions and oxidations occur at positions 3, 4, 5, and 17 as well as at C-11. A number of metabolic products can be isolated from the urine as well as from the plasma. The major urinary contributions come from dehydroepiandrosterone (which is also partially metabolized to androsterone and etiocholanolone) and the two major testosterone metabolites, etiocholanolone and androsterone.

ANDROSTERONE
(3α-HYDROXY-5α-ANDROSTAN-17-ONE)

ETIOCHOLANOLONE
(3α-HYDROXY-5β-ANDROSTAN-17-ONE)

There is also a small amount of epiandrosterone present. From the metabolism of adrenosterone, as well as the corticosteroids, some 11-oxygenated etiocholanolone and androsterone derivatives may be found in normal urine. In the process of synthesis and/or utilization of the androgenic substances, conjugation either with sulfuric acid or glucosiduronic acid also occurs. According to the report by Wotiz et al. [1], approximately 40% of urinary androsterone appears as the glucosiduronate while the rest is sulfate-conjugated. Roughly, the reverse ratios are found for the excretion of etiocholanolone, while dehydroepiandrosterone appears almost exclusively as the sulfate. The $C_{19}O_3$ steroids appear primarily as the glucosiduronates.

While androgens as such are generally determined by bioassay, their metabolic products, the 17-ketosteroids, are usually determined by means of the Zimmermann color reaction. Since this presents the investigator with a rather limited view of the metabolic picture, a number of useful methods have been described which allow the fractionation of the individual components and usually measure from five to eight individual compounds. Fractionation can be achieved either by utilization of adsorption columns [2, 3], gradient elution chromatography [4, 5], or the rigorous application of one or more paper chromatographic systems [6, 7]. The greatest drawback to any of these procedures is the number of manipulations which require considerable time.

Relatively few methods have been described for the plasma determination of these steroids. Noteworthy among these are the techniques described by Gardner [8] and Migeon and Plager [9], who presented evidence for the isolation from plasma of dehydroepiandrosterone,

androsterone, and etiocholanolone. At least two other components may be detected in normal plasma: free testosterone and testosterone glucosiduronate.

EVALUATION OF METHODS

A number of laboratories have reported the separation of several 17-ketosteroids using gas chromatographic procedures and some evidence has been presented for the usefulness of this technique in application to the study of the urinary 17-ketosteroids. One of the more specific, but rather involved, procedures has been described by Sparagana, Mason, and Keutmann [10], which was based in part on an earlier report by Haahti et al. [11], who showed the presence of peaks corresponding to the three major 17-ketosteroids in urine extracts. The difficulty with both of these procedures is based on the fact that only two of the three major urinary metabolites may be separated and at least two columns are necessary for quantitative estimation requiring the measurement first of compounds 1 and 3, and then of compounds 2 and 3, individual values being derived by subtraction. Complete separation of the major $C_{19}O_2$ and $C_{19}O_3$ 17-ketosteroids has been reported by Hartman and Wotiz [12], who separated the trimethylsilyl ethers of the $C_{19}O_2$ and $C_{19}O_3$ steroids on a silicone nitrile elastomer (XE-60) and on Hi-Eff 8B, a cyclohexane dimethanol succinate (CDMS) polymer. VandenHeuvel et al. [13] reported similar findings somewhat earlier for the $C_{19}O_2$ steroids only, using several different stationary phases, including the silicone nitrile (CNSi) as well as neopentylglycol succinate (NGS), QF-1 (a fluoroalkyl silicone elastomer), and SE-30. Adequate separation of the TMSi ethers appeared to be obtained only on the nitrile and the polyester phases. Furthermore, since neutral extracts would be expected to contain pregnanediol, both groups of workers showed the separation of this substance from the 17-ketosteroids. More recently, Creech [13a] has described a separation not only of the $C_{19}O_2$ but also of the $C_{19}O_3$ steroids and the progesterone metabolites. Unfortunately, no biological data are as yet available for this procedure. Reproducibility of the gas chromatographic procedure using the approach described by VandenHeuvel has been studied by France et al. [14], who showed that quantitative analysis down to the level of $0.05\,\mu g$ may be feasible. In a series of injections of androsterone, etiocholanolone, and dehydroepiandrosterone, the variation for duplicate samples was 1.5, 4.5, and 1.8% at the $0.3\,\mu g$ level. This work has recently been expanded to show its application to a routine determination of urinary $C_{19}O_2$ 17-ketosteroids [15]. The procedure includes both enzymatic and solvolytic cleavage of the conjugates, several washings, silica gel column chromatography, and a rather cautious derivative forming procedure.

A more thoroughly studied procedure for the determination of the 17-ketosteroids using gas chromatography is the one described by Kirschner and Lipsett [16]. Following hydrolysis (a subject to be discussed later) and ether extraction of the urine, the neutral portion of the extract is separated by thin-layer chromatography and divided into four major fractions. That portion of the chromatogram known to contain the 17-ketosteroids is then scraped off, eluted, converted to the trimethylsilyl ethers, and gas chromatographed on an XE-60 column.

By careful application of the procedures outlined, Kirschner and Lipsett were able to obtain sufficient statistical data to indicate that this is a very useful procedure in the analysis of 17-ketosteroid fractions. Although no specific studies were carried out to determine the sensitivity, the authors' experience suggests that as little as $50 \mu g$ per 24 hr of the compounds involved can be analyzed with confidence in a urine sample. In order to study the accuracy of the method, 1 mg of androsterone and 0.5 mg each of dehydroepiandrosterone, 11-ketoetiocholanolone, and 11β-hydroxyandrosterone were added to several 24 hr urine collections. In all instances excellent recovery was achieved. Precision of the method is somewhat more variable, apparently depending on the nature of the steroid. For androsterone, etiocholanolone, and dehydroepiandrosterone, when chromatographed on a QF-1 column, coefficients of variation were 4.7%, 4.0%, and 5.1%, respectively. Somewhat greater variations were found in the measurement of 11β-hydroxyetiocholanolone (7%), and 11-keto-etiocholanolone (10%) on an XE-60 column. These latter measurements were made at levels of 0.45 and 0.22 mg per 24 hr, respectively.

Evidence for the specificity of the reaction is based primarily on the preliminary separation of the 17-ketosteroids by thin-layer chromatography in a benzene:ethyl acetate system, followed by gas–liquid chromatography and mixed gas–liquid chromatography on XE-60. The R_f values for a variety of steroids are shown in Table 15-1. Of the various steroids listed only pregnanediol and allopregnanediol would be likely to present some difficulty in GLC separation from the $C_{19}O_2$ steroids. However, this presents no problem, since these two groups of steroids are previously separated by TLC separation. On the other hand, the $C_{19}O_3$ compounds, which do not generally separate well from the pregnanediols during TLC, are resolved satisfactorily by gas chromatography as the trimethylsilyl ethers. Relative retention values are shown in Table 15-1.

Figures 15-1 and 15-2 show typical chromatograms obtained from urinary extracts for the $C_{19}O_2$ and $C_{19}O_3$ steroids as described by Kirschner and Lipsett. Normal values for the various 17-ketosteroids in a series of males and females were determined by these investigators (Table 15-2).

As was mentioned earlier, here again the investigator is faced with the choice of speed by direct gas chromatographic analysis or a more elaborate preliminary purification to ensure greater specificity. While the latter is of course quite desirable, in clinical analysis the need frequently arises to obtain results in a short span of time and with the least expense to the patient.   In the event that such rapid analysis is necessary, the method described by Hartman and Wotiz [12], which relies entirely on the high resolution of the gas chromatographic column,   may well be applied to extracts following some form of relatively fast hydrolysis, such as that reported by Vestergaard [17] or DePaoli et al. [18].   Table 15-3 shows that separation of all the important 17-ketosteroids can be achieved as the trimethylsilyl ethers by chromatography on an XE-60 column.   Figure 15-3 shows the

TABLE 15-1. R_f Values* and Relative Retention of Some Neutral Steroid Metabolites†

Steroid	R_f	SE-30 1%	NGS 1%	XE-60 2%
$C_{19}O_2$:				
Androsterone	0.47	0.34	1.44	1.56
Etiocholanolone	0.41	0.32	1.87	1.81
Dehydroepiandrosterone	0.47	0.34	2.20	2.20
Androstan-3β-ol-17-one	—	0.34	2.20	2.23
Etiocholan-3β-ol-17-one	—	0.33	1.49	1.58
Androstenediol	0.39	0.37	—	0.71
$C_{19}O_3$:				
11-Ketoetiocholanolone	0.28	0.40	4.15	4.10
11β-Hydroxyandrosterone	0.30	0.61	5.25	4.53
11β-Hydroxyetiocholanolone	0.26	0.55	6.90	5.40
$C_{21}O_2$:				
Pregnanediol	0.29	0.62	1.85	1.53
Allopregnanediol	0.33	0.63	—	1.77
Pregnanolone	0.47	0.56	—	2.52
Pregnenediol	0.33	0.65	—	1.77
Pregnenolone	0.47	0.59	—	3.12
$C_{21}O_3$:				
Pregnanetriol	0.12	0.96	—	3.75
17-Hydroxypregnanolone	0.29	0.78	—	5.20
Pregnenetriol	0.28	1.03	3.30	3.73
17-Hydroxypregnenolone	0.37	0.83	—	6.40

*System: benzene: ethyl acetate (40:60).
†Reproduced by permission from M. A. Kirschner and M. B. Lipsett, Steroids 3:283 (1964).

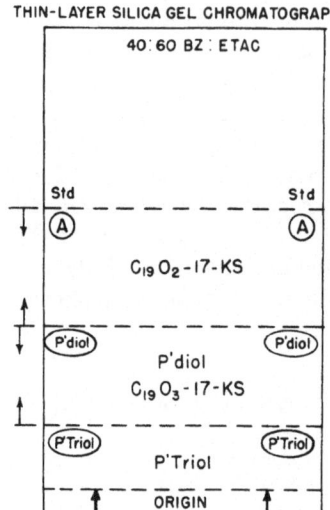

THIN-LAYER SILICA GEL CHROMATOGRAPHY

Figure 15-1. TLC of urinary 17-keto-steroids. Thin-layer chromatogram representing several urinary steroid fractions in benzene:ethyl acetate (40:60) system. The areas between the arrows at left are scraped off individually and eluted for further GLC (Figure 15-2). [Kirschner and Lipsett, Steroids 3:277 (1964)].

Figure 15-2. GLC of urinary 17-keto-steroid TMSi ethers. Typical chromatograms obtained for the two areas representing (A) the $C_{19}O_2$, and (B) the $C_{19}O_3$ fractions from the TLC fractionation of a urine extract (Figure 15-1). [Kirschner and Lipsett, Steroids 3:277 (1964)].

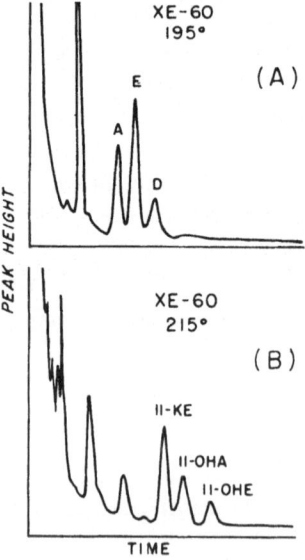

TABLE 15-2 17-KS Determination in Normal Human Subjects*

Steroid	Sex	No. of subjects	Mean	Range
Androsterone	Male	7	2.6	2.1-3.8
	Female	9	1.4	0.9-2.6
Etiocholanolone	Male	7	2.5	1.4-5.4
	Female	9	1.8	0.6-2.2
Dehydroepiandrosterone	Male	4	0.72	0.20-1.71
	Female	6	0.56	0.20-1.13
11-Ketoetiocholanolone	Male	7	0.54	0.42-1.41
	Female	9	0.48	0.25-0.85
11β-Hydroxyetiocholanolone	Male	7	0.39	0.22-0.67
	Female	9	0.43	0.10-1.18
11β-Hydroxyandrosterone	Male	7	0.46	0.16-0.89
	Female	9	0.27	0.05-0.48

*Reproduced by permission from M. A. Kirschner and M. B. Lipsett, Steroids 3:3 (1964), Table 5. Values for individual 17-KS ($C_{19}O_2$ and $C_{19}O_3$) as obtained by Kirschner and Lipsett using TLC and GLC.

TABLE 15-3. Separation of 17-KS and Pregnanediol by GLC on Several Stationary Phases*

Steroids (as TMSi ethers)	Relative retentions			
	XE-60†	SE-30‡	NGSeb§	Hi-Eff 8B¶
3α-Hydroxy-5α-androstan-17-one	1.02	0.47	0.60	0.78
3α-Hydroxy-5β-androstan-17-one	1.18	0.50	0.73	0.98
3β-Hydroxy-androst-5-en-17-one	1.36	0.56	0.84	1.14
3α-Hydroxy-5β-androstane-11,17-dione	2.40		1.36	2.03
3α,11β-Dihydroxyandrostan-17-one	2.57		1.87	2.59
5β-Pregnane-3α,20α-diol	0.78		0.70	0.68
3α-Hydroxy-5β-pregnan-20-one	1.44		1.08	1.38

*Reproduced (in part) by permission from I.S. Hartman and H.H. Wotiz, Steroids 1:33-38 (1963), Table I.
†Cholestane 9 min. 3% XE-60 on 80-100 mesh Diatoport S, 6 ft by $\frac{1}{8}$ in. (OD) stainless steel column. Column temperature 262°, pressure 14 p.s.i.
‡Cholestane 13.6 min. 1% SE-30 on 80-90 mesh Anakrom ABS, 6 ft by $\frac{1}{8}$ in. (OD) stainless steel column. Column temperature 251°, pressure 14 p.s.i.
§Cholestane 10 min. 1% Neopentylglycol sebacate on 80-90 mesh Anakrom ABS, 6 ft by $\frac{1}{8}$ in. (OD). Column temperature 231°, pressure 30 p.s.i.
¶Cholestane 6.3 min. 1% Hi-Eff 8B on 80-100 mesh Diatoport S, 6 ft by $\frac{1}{8}$ in. (OD) stainless steel column. Column temperature 242°, pressure 33 p.s.i.

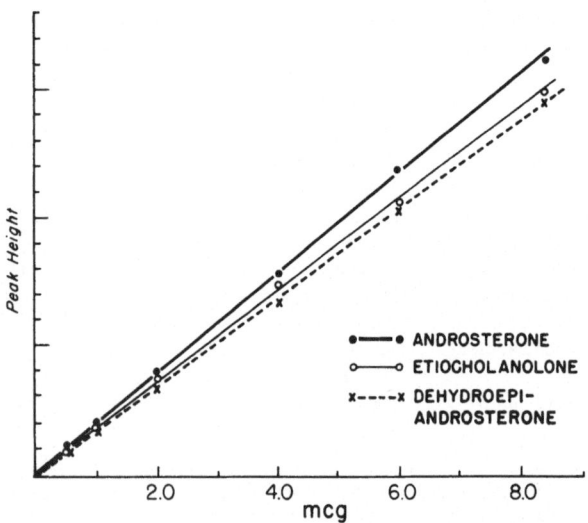

Figure 15-3. Response of the flame detector to varying amounts of $C_{19}O_2$ 17-ketosteroids.

linear response of three $C_{19}O_2$ 17-ketosteroid trimethylsilyl ethers. Figure 15-4 shows a chromatogram of a urine extract on a narrow bore, 12-ft glass column.

As a result of the Airlie House Conference on Gas Chromatography of Steroids, we have become aware of a very interesting study conducted at the Imperial Cancer Research Fund by B. S. Thomas and his colleagues. In order to develop not only a fast but also highly specific and precise method for the analysis of $C_{19}O_2$ 17-ketosteroids, these investigators compared separations of extracts on different gas chromatographic columns with those of their well-standardized gradient elution chromatographic method. It soon became apparent that there was a consistent overestimation of approximately 20% in the androsterone fraction. This overestimate was encountered whether XE-60 or Hi-Eff 8B columns were utilized and was found whether the method of Thomas and Bulbrook [19] for the preliminary extraction and purification or that described by Kirschner and Lipsett [16] was applied. Interposition of a Girard T separation, followed by GLC of the nonketonic fraction, showed the appearance of a peak following TMSi ether formation corresponding in retention time to that of androsterone. These investigators have resolved this unknown peak from androsterone by using a mixed column prepared by coating Chromosorb W (100-120 mesh) with 0.6% each of Hi-Eff 8B and the nonselective JXR silicone elastomer. (Gradient elution chromatography and GLC using the mixed column yielded values reasonably in agreement.)

Interestingly, it appears that in the GLC assay slightly lower values for etiocholanolone are found when compared to those from gradient elution chromatography. This suggests either a unique loss of etiocholanolone not observed for the other 17-ketosteroids or some lack of specificity of the gradient elution method.

In view of these findings, one must be cautious in applying any of the presently available GLC methods for the fractional analysis of 17-ketosteroids. All of those presently reported, with the exception of that of Creech et al. [13], which utilizes NGS columns, are based on separation with XE-60 or Hi-Eff 8B. Nevertheless, in a number of clinical situations it may well be reasonable to apply one of the methods described here, keeping in mind, however, the fact that a serious overestimate of the androsterone fraction will result.

It is expected that this elegant piece of research by Thomas will soon be followed by the description of a completely specific and reasonably rapid procedure.

Hydrolysis of Urinary 17-Ketosteroids

As was discussed earlier in this chapter, the 17-ketosteroids are generally excreted in the urine as a mixture of glucosiduronates and sulfates. In order to measure the different steroids, it is necessary first to liberate the steroid from the conjugate moiety so that it may

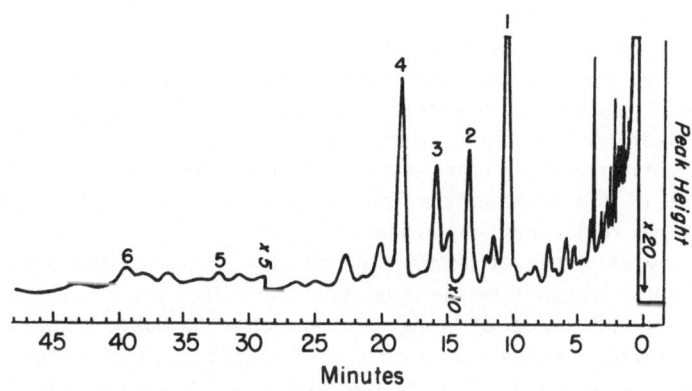

Figure 15-4. Analysis of 17-ketosteroids from a urine specimen. Column: 12 ft × 0.075 in. ID (glass); Column packing: XE-60 on specially treated and silanized Gas Chrom P 80-100 mesh support (4/96, w/w). Sample volume: 1 μl. Carrier gas (He) flow rate at column outlet: 30 ml/min. Temperature of injection port, column, and detector: 300, 200, and 240°C, respectively. Identified peaks: (1) 5β-pregnane-3α, 20α-diol (pregnanediol); (2) 3α-hydroxy-5α-androstan-17-one (androsterone); (3) 3α-hydroxy-5β-androstan-17-one (etiocholanolone); (4) 3β-hydroxy-androst-5-en-17-one (dehydroepiandrosterone); (5) 3α-hydroxy-5β-androstane-11,17-dione (11-keto etiocholanolone); (6) 3α, 11β-dihydroxy-5α-androstan-17-one (11β-hydroxy androsterone). (Courtesy of the Perkin-Elmer Corporation, Norwalk, Connecticut.)

become lipid-soluble. It has been known for some time, however, that the treatment of dehydroepiandrosterone sulfate with hot mineral acid causes destruction of this steroid.

A number of investigators have described procedures for the apparently complete and safe hydrolysis of these steroid conjugates. A rigorous discussion of the merits of one method over another is hardly part of this book. By and large, four different methods may be utilized.

One method described by Vestergaard and Claussen [17] requires relatively mild acid hydrolysis and simultaneous benzene extraction. This produces minimal destruction of dehydroepiandrosterone. Another procedure [21] takes advantage of the enzymatic cleavage which occurs on incubation of preparations from *Helix pomatia* or *Patella vulgata* with the steroid conjugates. This procedure, though mild, has two distinct disadvantages. First, it is time-consuming, taking about 24–48 hr, and second, the sulfatase content of this preparation is specific for the hydrolysis of 5β-sulfates only. Consequently, androsterone sulfate is not hydrolyzed and will not become lipid-soluble. In some instances this may be only a minor evil. The third method [20], which allows for virtually complete hydrolysis of all the conjugates, involves first the enzymatic liberation of steroid from glucuronic acid, followed by extraction of the free steroids and further treatment by continuous extraction with ether or other solvolytic reagents in the presence of acid, as described by Burstein. While this allows complete extraction of the 17-ketosteroids in an unaltered state, it is also the most time-consuming procedure.

Finally, the simultaneous nonenzymatic hydrolysis of most of the steroid conjugates has been reported by DePaoli et al. [18], who based their work on an earlier report by Burstein. The latter was able to show that when ethyl acetate or tetrahydrofuran is used in the presence of some perchloric acid, the sulfates and the glucosiduronates of steroids could be hydrolyzed and extracted. DePaoli compared several types of hydrolysis and concluded that the last mentioned would probably be the most suitable method for general usage. It has the advantage of being nondestructive to dehydroepiandrosterone sulfate, it appears to cause no artifact formation, it is relatively fast (approximately 3 hr), and it produces hydrolysis of most of the conjugates. These methods of hydrolysis, all of which are considered to be of interest to different investigators, are described below.

METHODS OF HYDROLYSIS
Method 1: Mild Acid Hydrolysis (Vestergaard and Claussen)
Thirty milliliters of urine are brought to pH 0.8 using 40% sulfuric acid. The mixture is transferred to a round-bottom flask and a

Friedrichs condenser is attached. An equal volume of benzene is added along with a few boiling chips. The biphasic solution is now rapidly brought to boiling with an electric heating mantle and refluxing is continued for 15 min. The flask is cooled immediately in running tap water, the upper (benzene) layer is withdrawn, and the pH of the urine is checked. Another volume of benzene, equal to the urine volume, is added, and refluxing is continued rapidly for 30 min more. Following this the urine is again cooled under running water and the benzene layer removed. At this point the solution is brought to pH 0.2 and after the addition of another equal volume of benzene, refluxing is carried out for 1 hr. The flask is cooled under running water as before and the benzene layer is separated. All the extracts are then combined and are ready for further treatment.

Method 2: Enzymatic Hydrolysis (Bush)

One-twentieth of a 24 hr specimen of urine is brought to pH 4.8−5.0 and made 0.5 M in acetate. The urine is now heated on a boiling water bath for 20 min, and after cooling sufficient enzyme (Glusulase-Endo Products; Helicase-Fisher) to obtain a concentration of 2000 units of β-glucuronidase per ml of urine is added. The mixture is then incubated at 37°C for 24 hr. The steroids are extracted twice with three volumes of a mixture of ether:ethyl acetate (2:1). The combined extracts are then ready for further partitioning.

Method 3: Combined Enzyme Hydrolysis and Solvolysis (Segal et al.)

A volume equal to 5% of the urine is adjusted to pH 4.7 with acetic acid or sodium hydroxide, as required. To the solution $\frac{1}{10}$ volume of acetate buffer (pH 4.7) is added. β-Glucuronidase is then added to produce a final concentration of 1000 IU/ml of urine, and the mixture is allowed to incubate at 40°C for 16 hr. The free steroids are extracted with 5 volumes of a mixture of petroleum ether:benzene (1:1). The urine is then neutralized with aqueous sodium carbonate solution and the mixture made 3 M with sodium chloride. The steroid sulfates are extracted with 2 volumes of ethyl acetate. The organic phase is dried with anhydrous sodium sulfate (6 g per 100 cc solution) and, following filtration, is evaporated to dryness. Sufficient sulfuric acid in ethyl acetate is added to the dried residue to make the final reaction mixture 2×10^{-4} M in sulfuric acid. After this solution is kept for 20 hr at 30°C, it is extracted with 100 ml of aqueous 1 N NaOH and twice with 100 ml of water. The solutions containing free steroid are combined and dried over sodium sulfate, at which point they are ready for preparation for GLC.

Method 4: Solvolysis in the Presence of HClO$_4$

(DePaoli, Nishizawa, and Eik-Nes)

Five percent of the urine specimen is brought to pH 4 and 50 g of sodium sulfate per 100 ml of urine is added. The steroid conjugates are extracted once with 3 volumes of ethyl acetate. After the ethyl acetate extract is made 0.01 M with perchloric acid, the mixture is allowed to stand at room temperature for 3 hr. Following this period the solution is washed twice with a $\frac{1}{10}$ volume of 10% potassium hydroxide and with $\frac{1}{10}$ volume of water. The ethyl acetate is dried over sodium sulfate, filtered and evaporated to dryness. After the residue is dissolved in 3 ml of ethanol containing 1% potassium hydroxide, it is left standing at room temperature for 1 hr. Next, the alcoholic solution is diluted with water and the free steroids extracted with a mixture of petroleum ether:benzene (1:1).

EXPERIMENTAL

Method A-I (Hartman and Wotiz)

As was pointed out earlier, the choice of hydrolytic procedure must rest with the individual investigator and depends on the nature of the information which is desired. The method of Vestergaard and Claussen or of DePaoli et al. appears to us to be entirely useful to prepare extracts well suited for simple gas chromatographic separation.

When hydrolysis of the steroid conjugates has been accomplished, the extract is washed with 25 ml of 1 N NaOH (not needed for methods 3 and 4) followed by two 25 ml water washes. The organic layer is dried over a little anhydrous sodium sulfate and the mixture is filtered. Following evaporation of the filtrate on a rotating evaporator, the residue is transferred to a 15 ml centrifuge tube with acetone. This solvent is evaporated under air flow and the remaining d r y material is dissolved in 1 ml of chloroform, and 1 ml of hexamethyldisilazane and 0.1 ml of trimethylchlorosilane is added. The well-stoppered tubes are heated to 60°C for 1 hr, after which time the solvent is evaporated in a stream of dry air. (This operation must be carried out in a fume hood and all steps hence require conditions of complete dryness since the TMSi ether cleaves spontaneously in the presence of water.) The residue is triturated with 5 ml of hexane and the mixture is centrifuged. The supernatant fluid is transferred to a small vial and the solvent is again evaporated. The thoroughly dry residue is finally dissolved in 0.1 ml of hexane to be used for GLC analysis.

A 6-ft, $\frac{1}{8}$-in.-OD column containing 3% XE-60 silicone nitrile elastomer on 80-100 mesh diatomaceous earth is prepared and cured for 48 hr, at 250°C, with a slow flow of nitrogen. After the relative peak heights of the different steroids to be analyzed have been determined, it is usually only necessary to determine the peak height of a single standard since the peak height ratios of the several components to be analyzed remain fairly constant.

Following determination of the peak height of a standard, a $1 \mu l$ aliquot of the hexane solution from the urine extract is injected into the column which is kept at 200°C with a flow rate of 80 ml/min.

Similar results may also be achieved by the use of the heat-stable polyester Hi-Eff 8B. This stationary phase is remarkably stable, but because of the close proximity of elution of androsterone and pregnanediol it should be used with caution when large amounts of the latter are expected, as in pregnancy.

Quantitative measurement may be made by comparing the height of each unknown peak to a curve prepared with a known concentration of authentic standard.

Method A-II (Kirschner and Lipsett)

Following hydrolysis of the 17-ketosteroid conjugates and their extraction, the solvents are evaporated and the dried residue is taken up in a little acetone. This solution is applied across 3–5 cm on a 0.375-mm silica gel plate.* The bulk of the residue determines the length of the line of application. At least 2 samples generally can be applied to a standard 20 × 20 cm plate. Twenty-five micrograms of androsterone, pregnanediol, and pregnanetriol are applied at each side of the plate. The chromatogram is developed for 45 min in the system, benzene: ethyl acetate (40:60). After drying in air, the plate is sprayed with 0.1% of rhodamine 6G in ethanol. The plate is then examined under ultraviolet light where the nonabsorbing steroids appear as white lines on the yellow background.

Three zones are outlined: Zone 1 includes the area from just above the androsterone standard to the pregnanediol standard, zone II includes the area from just above the pregnanediol standard to the pregnanetriol standard, and zone III includes the area from just above the pregnanetriol standard to the line of application. Zones I and II are scraped off the plate into 15 ml centrifuge tubes, eluted twice with 5 ml acetone, and dried under an air stream in 8 ml teflon-capped vials.

Formation of Trimethylsilyl Ethers. The dried eluates are dissolved in 0.2 ml of chloroform to which is added 0.2 ml of hexamethyl-disilazane and 10 drops of trimethylchlorosilane. The vial is capped with a tight teflon seal and placed in a water bath at 56° for $\frac{1}{2}$ hr, or at room temperature overnight. The powdery white precipitate which forms on adding trimethylchlorosilane does not interfere with the subsequent measurement. Before analysis by GLC, the excess

---

*Silica Gel G, Brinkman Co., Great Neck, N. Y., or Adsorbosil 1, Applied Science Laboratories, Inc., State College, Pa.

reagents are blown off under a stream of dried air, and the steroid TMSi ethers taken up in hexane to a final concentration of approximately $1\mu g/\mu l$. 5α-Androstan-17-one is used as a marker for relative retentions and is taken up into the syringe just before injection into the column.

Gas-Liquid Chromatography.   A 6-ft-long × 3.4-mm-ID glass column is packed with Gas Chrom P (60-80 mesh), coated with 2% XE-60. The carrier gas input pressure is set to allow a flow of ~25 ml/min. The flash heater and detector are kept about 50° higher than the column temperature, which ranges from 195° for the $C_{19}O_2$ metabolites to 215° for the $C_{19}O_3$ steroids.

$C_{19}O_2$ — 17 KS.   The TMSi ethers from zone I are analyzed by GLC using a 2% XE-60 column operating at 195° and 20 psi. In this system, androsterone, etiocholanolone, and dehydroepiandrosterone appear as separate peaks (Figure 15-2a), and each is quantitated by comparing the area response with known amounts of appropriate standards.

$C_{19}O_3$ — 17 KS.   The steroids in zone II are similarly eluted with acetone and converted into TMSi ethers.   These $C_{19}O_3$ –17 KS– TMSi ethers are separated and measured on GLC using the phase 2% XE-60 at 215° and 20 psi (Figure 15-2b).

DETERMINATION OF TESTOSTERONE IN URINE

Evaluation of Methods

Although the determination of the individual 17-ketosteroid fractions can now be achieved in a relatively short time and with considerably less effort than needed previously, there still remains the question of the significance of this particular assay in the differentiation of disease.   One of the difficulties in correlating the increase or decrease of any of these fractions with disease states relates of course to the fact that the precursors to the 17-ketosteroids are secreted by more than one gland. On the other hand, the sole major contributor to this steroid pool by the testis is testosterone. It is therefore considered by some to be more important to analyze the urine or plasma concentration of testosterone than its metabolites. For instance, as Dorfman [22] has pointed out, virilization in women may well be accompanied with normal values for the individual 17-ketosteroids since apparently the prime responsibility for the virilizing effect lies with a relatively small amount of testosterone formed in the ovary or adrenal cortex. Its very high biological activity may therefore cause profound biologic changes without significantly altering the total excretion or relative proportion of the individual 17-ketosteroids.

The first application of gas–liquid chromatography to the urinary analysis of testosterone was described by Futterweit et al. in 1963 [23]. These workers were able to measure testosterone at a concentration of as low as $15 \mu g$ per 24 hr with a recovery varying from 67–98% of added steroid. The precision of the method is reported as ±7%, which appears to be quite acceptable at the levels of measurement. The method itself involves hydrolysis and preparation of a neutral extract, which is then followed by Girard separation of the ketonic fraction. Following thin-layer chromatography, the testosterone fraction is eluted and chromatographed on a short SE-30 column with a relatively high concentration of stationary phase (6.8%). Values for normal males in different age groups are reported as varying from a low of $28 \mu g$ for an individual above 55 years of age to $250 \mu g$ for a young man. This appears to be in reasonable agreement with other procedures not utilizing gas–liquid chromatography. Values of 2.4 to $8 \mu g$ of testosterone are reported for females, but one would question these data on the basis that the lower limit of sensitivity is $15 \mu g$ per 24 hr. Although the paper does not give a typical chromatogram, the authors' description that the steroid is generally found appearing as a shoulder on a much larger peak (one of the major 17-ketosteroids) suggests that a different form of preliminary purification would be more desirable.

Indeed, four methods have been published recently which use gas–liquid chromatography but apply different methods of preliminary purification, thus alleviating the particular difficulties encountered by Futterweit et al. Values obtained using these four procedures, with the method of Futterweit et al. and with classical methods, show rather similar excretion levels of testosterone in human urine.

Early in 1964, Ibayashi et al. [24] described a procedure involving β-glucuronidase hydrolysis of the testosterone glucosiduronate, followed consecutively by preparation of a neutral extract, thin-layer chromatography, acetylation, and further thin-layer chromatography of testosterone acetate. Final separation and analysis was carried out on a $4\frac{1}{2}$-ft, 1.5% SE-30 column.

Even though Ibayashi et al. neglected to show a typical chromatogram from a urine extract, it is presumed that, in the absence of any mention of interfering substances, a relatively clean chromatogram may be obtained in this manner. Replicate analyses after the addition of steroid to urine yielded excellent results, with values of 100.4 ± 0.64 μg recorded. Unfortunately, no precision data at low levels of excretion were presented. The lower limit of sensitivity appears to be 5 μg per 24 hr with an average recovery of 55% of added steroid.

Mean values for all males (between the ages of 18 and 42) are reported as 131 ± 15 μg and levels for several females were from

TABLE 15-4. R_f Values of Steroids on TLC of Kieselgel HF_{254}[*]

Steroid	Benzene: ethyl acetate (1:1)	Benzene: ethyl acetate (3:1)
Progesterone	0.52	0.31
Androsterone	0.43	0.20
Etiocholanolone	0.37	0.15
Testosterone	0.30	0.12
Dehydroepiandrosterone	0.40	0.22
4-Androstene-3,17-dione	0.44	0.24
Corticosterone	0.15	0.03
Cortisol	0.11	0.03
Testosterone acetate	0.57	0.38
DHEA acetate	0.67	0.56
Androsterone acetate	0.67	0.56
Etiocholanolone acetate	0.67	0.56
5-Androstene-3β,17β-diol	0.70	0.60

*Reproduced by permission from H. Ibayashi, M. Nakamura, S. Murakawa, T. Uchikawa, T. Tanioka, and K. Nakao, Steroids 3:559 (1964).

less than 5 to 8 μg. Subjects in their twenties excreted up to 200 μg, while normal subjects in their sixties and seventies excreted less than 25μg of testosterone per 24 hr. For females, values of less than 5μg per 24 hr were found in age groupings beginning at twenty and ending at seventy years. Considerably elevated values were detected in Stein–Leventhal syndrome (319 μg per 24 hr), in hirsutism, and in three cases of Cushing's syndrome. The specificity of the method is based on the two thin-layer chromatographic separations (Table 15-4), on the retention time of the steroid acetate during gas–liquid chromatography (Table 15-5), and on the appearance of symmetrical peaks with exact retention times and increased peak heights following the addition of small concentrations of testosterone to the extract.

Shortly thereafter, Brooks [25] published the first procedure which allowed the simultaneous determination of testosterone and epi-testosterone in urine. This method involved the usual hydrolysis and extraction followed by thin-layer chromatography, acetylation, further thin-layer chromatography of the steroid acetates, and finally gas–liquid chromatography.

Recoveries for individual experiments were based on the recovery of added ^{14}C-testosterone. Two modifications were incorporated in the original description. In the event urine from children is analyzed, a Girard separation of the crude extract should be carried out before thin-layer chromatography is attempted.   Furthermore, in the event that testosterone production rates need to be determined, a paper chromatogram (ligroin–propylene glycol system) is interposed between the first TLC and acetylation. Recoveries of added radioactive testosterone following the complete procedure ranged from 39 to 66%. The authors determined the accuracy of measurement of both testosterone and epitestosterone for amounts of mixtures varying from 0.4 to 6 μg and the results are shown in Table 15-6. A typical series of chromatograms is shown in Figure 15-5.   The column used for this separation was a 1.5 m QF-1 column.   However, it would appear, from an examination of these tracings, that the separations of epitestostcrone acetate and testosterone acetate are rather minimal, raising some doubt as to the accuracy of quantitative determination of the peaks. Since other methods using separation techniques which are no more complicated, and gas chromatographic columns allowing greater resolution of these two peaks are now available, details of this method are not presented here. It should be pointed out, however, that the values for normals presented by these authors appear to be well within the numerical values established by a number of other procedures.

TABLE 15-5. Relative Retention of Steroids on
GLC (1.5% SE-30 on Anakrom A)*

Steroid	r
Cholestane	1.00 (9.5 min)
Androsterone	0.42
Etiocholanolone	0.42
Dehydroepiandrosterone	0.40
4-Androstene-3, 17-dione	0.59
Androsterone acetate	0.52
Etiocholanolone acetate	0.52
DHEA acetate	0.59
Testosterone	0.62
Testosterone acetate	0.81
5-Androstene-3β, 17β-diol	0.45
5-Androstene-3β, 17β diacetate	0.87
Progesterone	0.91
17α-OH-Progesterone	1.41
5-Pregnenolone	0.68
17α-OH-Pregnenolone	1.07

*Reproduced by permission from H. Ibayashi, M. Nakamura, S. Murakawa, T. Uchikawa, T. Tanioka, and K. Makao, Steroids 3:559 (1964).

TABLE 15-6. Error of Measurement of Known
Mixtures of Testosterone and Epitestosterone
Acetates by Gas Chromatography*

Ratio Epitestosterone acetate / Testosterone acetate applied to column	Error of measurement (%)	
	Testosterone	Epitestosterone
∞		− 1
10	−10	−18
5	+ 8	− 7
2.5	− 5	− 5
2	+ 4	− 7
1.25	+ 2	−12
1	+ 10, −1	−11, −9
0.5	+ 1	− 3
0	+ 1	

*Reproduced by permission from R.V. Brooks and G. Giuliani, Steroids 4:101 (1964).

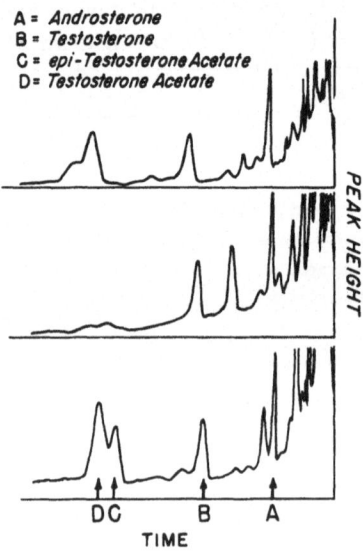

A = *Androsterone*
B = *Testosterone*
C = *epi-Testosterone Acetate*
D = *Testosterone Acetate*

Figure 15-5. Gas chromatography traces from three subjects [R. V. Brooks and G. Guiliani, Steroids 4:117 (1964)].

Late in 1964, Sandberg et al. presented a different approach to the analysis of urinary testosterone by incorporating essentially all of the early steps of the method of Camacho and Migeon [26], but instead of using a spectrometric assay of individual fractions, determined the testosterone concentration by gas–liquid chromatography. Recoveries using this procedure, which is considerably more tedious than the method of Ibayashi since it incorporates a Florisil chromatogram and a rather lengthy gradient elution chromatogram, are of the order of 90%. Relatively little statistical data are offered by the authors; however, considerable validity is implied by a comparison of the results obtained using both the complete and the GLC modified method of Camacho and Migeon on identical urine extracts. Quantitation is derived by the addition of epitestosterone as an internal standard during gas chromatographic separation on a 2% XE-60 column. The steroids were measured as the trimethylsilyl ethers.

Interestingly, in a number of instances lower excretion values were obtained following gas–liquid chromatography as compared to the spectrophotometric procedure (Table 15-7). This might imply that on occasion the preliminary purification is not entirely adequate and that further necessary purification does indeed occur on the gas–liquid chromatogram, imparting increased specificity to the GLC modified procedure. This latter point is further borne out by a typical chromatogram (Figure 15-6) of the combined testosterone zones of the gradient elution chromatogram which shows a number of unidentified early peaks.

The details of this method are not described here since it does not really take advantage of the efficiency of the GLC column, resulting in an unduly difficult separation procedure incorporating three different chromatographic systems. The contribution of this particular paper by Sandberg et al. [27] lies in the excellent work of comparing GLC with conventional methods. The fact that the data are in good agreement with those of Ibayashi et al. [24] engenders a greater feeling of security with the latter's simpler procedure.

Finally, just prior to completion of this manuscript, a method appeared for the simultaneous gas chromatographic analysis of urinary testosterone and epitestosterone [28]. This method, by inclusion of Δ^1-testosterone as an internal standard which is carried through all steps other than hydrolysis, eliminates the necessity of adding a radioactive standard to check recovery and allows essentially direct reading of corrected testosterone excretion data. The procedure, which is based on the method of Ibayashi, includes the thin-layer separation of the steroid followed by acetylation and oxidative purification. This in turn leads to another thin-layer chromat-

TABLE 15-7. Urinary Testosterone as Measured by Gas Chromatographic Techniques and Ultraviolet Spectrophotometry[*]

| | | | Exeretion of testosterone (μg per 24 hr) [†] | |
| | | | Gas chromatography after gradient elution | Ultraviolet absorption after paper chromatography |
Subjects	Age	Sex	Testosterone	T[‡]
1.	33	Male	55	57
2.	37	Male	401	441
3.	14	Male	180	180
4.	33	Male	330	337
5.	33	Male	66	52
6.	33	Male	53	70
7.	41	Male	79	99
8.	27	Male	73	76
9.	27	Male	122	128
10.	37	Male	57	78
11.	29	Male	48	67
12.	29	Male	47	80
13.	13	Male	4	8
14.	73	Male	18	20
15.	80	Male	16	23
16.	32	Male	112	132
17.	29	Male	138	140
18.	29	Male	140	142
19.	5	Male	0	0
20.	11	Male	0	0

[*]Reproduced by permission from D.H. Sandberg, N. Ahmad, W. Cleveland, and K. Savard, Steroids 4:557 (1964).
[†]The above values for testosterone excretion were obtained in various experimental conditions, including stimulation with human chorionic gonadotropin (subjects 2 and 4).
[‡]Corrected for losses of added testosterone-[14]C. Recoveries of added radioactive testosterone averaged 91% when estimated following gradient elution from the alumina column.

ogram of the derivatives and finally gas chromatography on an SE-30 column.

The author demonstrated the expected linear curve for testosterone acetate in a flame ionization detector and furthermore showed that there was a direct linear relationship between the responses of testosterone acetate, epitestosterone acetate, and Δ^1-testosterone acetate. Reasonably convincing evidence for the specificity of the assay technique is presented. It is based on a comparison of the retention times of the compounds in question from an extract with authentic standards. Augmentation of a urine extract presented no evidence of asymmetrical peak-spreading or peak separation. Further evidence was obtained by hydrolysis of the finally derived acetate, preparation of the trimethylsilyl ethers, and comparison of the retention times of testosterone TMSi ethers and the free compound with extract on SE-30 and XE-60 columns. Further evidence of specificity is derived from the two thin-layer chromatograms of the free and acetylated steroids.

The accuracy of the method was determined by adding known amounts of steroid to simulated urines and a normal urine specimen, followed by the complete analytical procedure. Absolute recovery of added steroid was 75%, with relative recoveries of testosterone and epitestrosterone, based on comparison with Δ^1-testosterone recovery, of 93–108%.

Estimation of the precision, based on 10 aliquots of the same urine, yielded coefficients of variation of 5% for testosterone and 9.2% for epitestosterone. The rather larger value of the latter is to be expected, of course, in view of the very low excretion levels of this steroid.

Sensitivity data are not yet available on this procedure. Theoretical values based on the sensitivity of pure compounds in the flame ioniza-

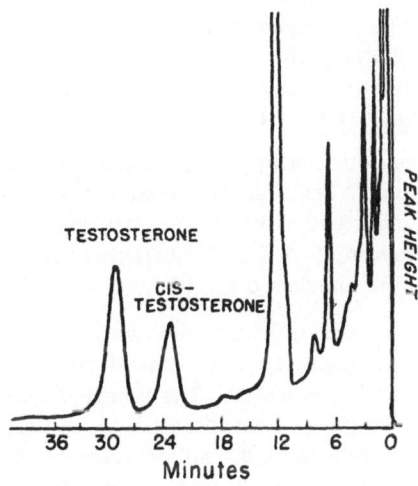

Figure 15-6. Testosterone in urine. After addition of cis-testosterone, the trimethylsilyl ethers were formed and chromatographed on 2% XE-60 as the stationary phase. [D Sandberg, Steroids 4:561, Figure 3 (1964)].

tion detector are of course not achievable because of interference from urinary contaminants. Only a few clinical data obtained by use of this method are as yet published, but the author states that the values of testosterone and epitestosterone obtained for a series of normal males are in good agreement with those obtained by a number of other investigators using both gas chromatographic and more conventional means of analysis.

EXPERIMENTAL

Method TU-I (Ibayashi et al.)

Materials. Beef Liver β-Glucuronidase. 10,500 Fishman units per ml.

Kieselgel HF$_{254}$. (E. Merck AG, Darmstadt, Germany). A slurry is prepared by mixing 30 g of Kieselgel with 80 ml of distilled water. The glass plates, coated to 0.25 mm thickness, are activated at 110°C for 30 min and transferred to a storage cabinet with desiccant.

1,2-^{3}H-Testosterone. (Specific activity 150 mC/mg.) Purified in a Bush A system.

Liquid Scintillation Fluid. Four grams of 2,5-diphenyloxazole (PPO) and 100 mg of 2,4-bis-[2-(5-phenyloxazolyl)]-benzene (POPOP) are added to 1 liter of toluene. Fifteen milliliters of this solution is added to the dried aliquot of purified extract.

Ethyl Ether. Peroxide free, analytical reagent.

Benzene. Reagent grade.

Ethyl Acetate. Reagent grade.

Methanol. Redistilled, reagent grade.

Pyridine. Reagent grade.

Acetic Anhydride. Reagent grade.

Hydrolysis. One-fourth aliquot of a 24 hr urine is adjusted to pH 5.0 with 50% acetic acid. Ten milliliters of 1 M acetate buffer (pH 5.0), 2 ml of chloroform, 500 units of β-glucuronidase per ml of urine, 0.1 g of streptomycin, and 1,2-^{3}H-testosterone (ca. 9000 cpm) are added and the mixture incubated for 96 hr at 37°C.

Extraction. The urine is extracted twice with 1 volume of ether. The combined ether extracts are washed twice with 0.1 volume of cold 1 N NaOH and distilled water. The ethereal extract is dried over anhydrous sodium sulfate, concentrated to dryness in a rotatory evaporator, transferred to a 30 ml centrifuge tube, and brought to dryness under a gentle stream of nitrogen at ~40°C.

Thin-Layer Chromatography. The urinary extract is applied at a 3 cm distance from the edge of a 20 × 20 cm glass plate coated with Kieselgel HF$_{254}$, while testosterone standards (ca. 10 μg) are applied near the edges.   The chromatogram is developed in benzene:ethyl acetate (1:1 v/v) until the solvent front reaches 15 cm. After development, the plates are dried at room temperature. Following detection of the spots of testosterone reference standards with an ultraviolet scanner (253.6 mμ), which makes 0.2 μg of testosterone detectable without reagent spray, the specimen zone corresponding to testosterone standard is carefully removed and thoroughly eluted with redistilled methanol. The eluate is evaporated under a nitrogen stream and 0.1 ml each of acetic anhydride and pyridine is added to the residue. The acetylation mixture is kept overnight at room temperature. The acetylated sample is again developed for about 35 min on a chromato-plate of Kieselgel HF$_{254}$ with benzene:ethyl acetate (3:1 v/v), using testosterone acetate as a reference standard (solvent front 15 cm). The zone of the urinary extract corresponding to the testosterone acetate standard is removed, thoroughly eluted with methanol, and dried in a stream of nitrogen for gas chromatographic analysis.

Gas-Liquid Chromatography.   The dried residue of the eluate is dissolved in 50 μl of acetone. A 5 μl portion is injected directly into a 1.5 m by 4 mm (ID) column filled with 1.5% SE-30 on Anakrom A (60-100 mesh).   The column is maintained at 216°C, with detector temperature of 240°C and flash heater temperature of 240°C. A flow rate of nitrogen is maintained at 30 ml/min. Quantification of the peak is carried out by comparing the testosterone acetate peak area of the sample with that of a standard of similar concentration, after having demonstrated a linear response of the detector in the specific range of concentration studied.   An aliquot is removed for counting in a liquid scintillation spectrometer employing 15 ml of a PPO–POPOP system.   The total amount of acetylated testosterone in a 24 hr urine specimen is calculated by allowing for the losses suffered during extraction, TLC, and acetylation as determined by tritium analysis. Finally, the values of urinary testosterone are recalculated on the basis of the ratio of molecular weights of testosterone acetate to testosterone.

Method TU-II (Sparagana)
Materials. Reagents. All reagents used are of analytical grade. Ethanol, methanol, benzene, and dichloromethane are distilled through a 12 in. fractionating column filled with glass helices. Ether is freshly distilled and sodium diethyldithiocarbamate added.

Methodology.   One-twelfth of a 24 hr urine is extracted with ether and the extract is discarded.   The urine is then adjusted to pH5.0 with 50% acetic acid. Then, 3 ml of 1 M acetate buffer (pH 5.0), 1 ml of chloroform, 500 units of β-glucuronidase per ml of urine, 0.05 g of streptomycin, and 20 μg of Δ^1-testosterone are added to the urine and the mixture is incubated for 96 hr at 37°C.

Extraction. Following incubation, the urine is extracted twice with an equal volume of ether. The combined ether extracts are washed twice with 0.1 volume of cold 1 N NaOH and distilled water. The ether extract is dried over anhydrous sodium sulfate and evaporated to dryness in a rotary still. The residue is transferred to a centrifuge tube of approximately 30 ml capacity and the solvent is evaporated under a gentle stream of nitrogen.

Thin-Layer Chromatography of Free Steroids. A thin-layer plate (0.5 mm thick) of Silica Gel GF$_{254}$ is prepared and the extract is applied to the center of this plate. On each side of the extract, and removed from it, spots of testosterone and Δ^1-testosterone are applied as comparison standards. The chromatogram is developed in a lined chamber with a mixture of 200 ml of dichloromethane and 20 ml of methanol. The solvent front is allowed to advance to a point 10 cm above the origin. An area of the extract bounded by the two standards, as determined by examination under ultraviolet light, is removed carefully with a spatula into a sintered glass funnel. Steroids are then eluted with two 15 ml portions of acetone.

Acetylation and Oxidation. The dried extract obtained is taken up in a mixture of 0.2 ml of pyridine and 0.1 ml of acetic anhydride and kept at room temperature for 16 hr. Next, the steroid acetates are partitioned between 10 ml of dichloromethane and 2 ml of 20% aqueous methanol. The dichloromethane is washed twice with 1 ml portions of water and then taken to dryness under nitrogen. Oxidation is achieved by adding 0.5 ml of a solution of chromic acid (20 mg per 10 ml of 36% acetic acid) to the dried residue and keeping the mixture at 45° for 1 hr. After extraction of the steroids with 10 ml of dichloromethane and washing the extracts with two 1 ml portions of water, the solvent is again evaporated.

Thin-Layer Chromatography of Steroid Acetates. The dried residue thus obtained is applied to another thin-layer plate as described above, but using Δ^1-testosterone acetate and testosterone acetate as the markers. The chromatogram is developed as before, but in a mixture of 200 ml of benzene and 25 ml of acetone. Elution of the steroid is carried out exactly as described earlier, by scraping the areas bounded by the steroid acetate standards into a sintered glass funnel and eluting with acetone.

Chromatography. Final separation and quantitation are achieved in a 12-ft, gas chromatographic column packed with Gas Chrom P (80-

100 mesh) which is acid washed, silanized, and then coated with 3% SE-30. The extract, which is transferred to a small conical centrifuge tube, is dissolved in 100 μl of acetone, and 3 μl of this solution is injected with a 10 μl Hamilton syringe. A typical chromatogram obtained is shown in Figure 15-7.

Tables 15-8 and 15-9 show R_f values for the two thin-layer chromatograms as well as retention relative to cholestane on SE-30.

Quantitative determination is carried out by peak area measurement and by inserting the proper values in the following equation:

$$X = \frac{iP_x}{mP_i}$$

where X is weight of testosterone, i is weight of Δ^1-testosterone added, P_x is peak area of testosterone acetate, P_i is peak area of Δ^1-testosterone acetate, and m is calibration constant from the curve.

A similar equation is used to calculate the concentration of epitestosterone.

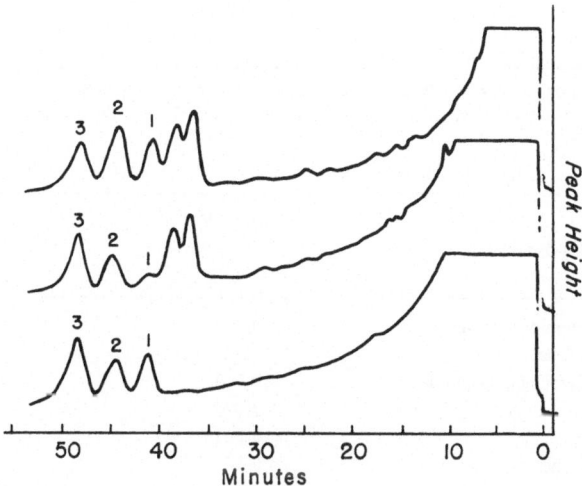

Figure 15-7. Gas chromatogram illustrating recovery of testosterone acetate and epitestosterone acetate from synthetic urine. Upper: Authentic urine with added steroid mixture.* Middle: Authentic urine with no added steroid. Lower: Steroid mixture* acetylated, direct injection. Peak areas: (1) Epitestosterone acetate. (2) Testosterone acetate. (3) Δ^1-Testosterone acetate. Conditions: 12 ft, 3% SE-30 column, 3 μl of 100 μl acetone solution injected, oven temperature 220°C, injection block 290°C, carrier gas (argon) flow 30 cc/min, no effluent split, recorder 5 mV full scale, attenuation 2×, [M. Sparagana, Steroids 5:772 (1965)]. Steroids 5:772 (1965)].

*Contains 10.1 μg testosterone, 10.1 μg epitestosterone, and 20.2 μg Δ^1 testosterone.

TABLE 15-8. R_f Values for Free Steroids*

Steroid	Thin-layer CH$_2$Cl$_2$/ CH$_3$OH (200:20) R_f	Gas chromatography SE-30 r
Cortol	0.054	
Cortolone	0.066	
Allotetrahydrocortisol	0.167	
Tetrahydrocortisol	0.186	
Cortisol	0.228	
Tetrahydrocortisone	0.229	
Tetrahydro compound S	0.246	
Pregnanediol	0.353	0.617
11β-Hydroxyetiocholanolone	0.354	0.567
11β-Hydroxyandrosterone	0.397	0.631
11-Ketoandrosterone	0.433	0.432
11-Ketoetiocholanolone	0.436	0.430
Δ^1-Testosterone	0.439	0.569
Testosterone	0.488	0.523
Epitestosterone	0.491	0.516
Etiocholanolone	0.492	0.352
Dehydroepiandrosterone	0.525	0.375
Androsterone	0.607	0.379
Cholestane		1.00†

*Reproduced by permission from M. Sparagana, Steroids 5:775 (1965).
†Cholestane time 66 min; 12 ft, 3% SE-30; column 220°C; injection block 290°C; carrier gas flow 30 ml/min.

TABLE 15-9. R_f Values for Steroid Acetates*

Steroid acetate	Thin-layer benzene:acetate (200:25)	Gas chromatography
11β-Hydroxyandrosterone	0.314	0.835
Δ^1-Testosterone	0.323	0.805
11β-Hydroxyetiocholanolone	0.334	0.802
Testosterone	0.374	0.736
Epitestosterone	0.361	0.681
11-Ketoetiocholanolone	0.444	0.614
11-Ketoandrosterone		0.612
Dehydroepiandrosterone	0.536	0.564
Androsterone	0.547	0.508
Etiocholanolone	0.556	0.504
Pregnanediol	0.640	1.285
Cholestane		1.00†

*Reproduced by permission from M. Sparagana, Steroids 5:775 (1965), Table 2.
†Cholestane time 64 min; 12 ft, 3% SE-30; column 220°C; injection block 290°C; carrier gas flow 30 ml/min.

DETERMINATION OF TESTOSTERONE IN PLASMA

Evaluation of Methods

In view of the fact that a number of investigators have reported normal 17-ketosteroid excretion in hypogonadal men (eunuchoids or castrates), Dorfman [22] already has pointed out the need to analyze testosterone in the blood. Several such procedures have been described in the past few years, primarily based on a method originally described by Finkelstein in 1961 [29]. This is an extremely sensitive method which is dependent on the conversion of testosterone to estradiol through incubation with a placental enzyme preparation as described by Ryan [30], following a rather laborious isolation procedure for testosterone. After conversion, the mixture of estrone and estradiol obtained is again chromatographically separated and finally analyzed by fluorometry. Despite the obvious elegance of this procedure, it has found little application in clinical usage because of its considerable tedium. With the advent of the highly sensitive gas chromatographic detection devices, the picture appears to be changing and in the last two years two methods have been described for the analysis of plasma testosterone.

The first of these by Guerra-Garcia et al. [31] has the advantage of considerable speed but presently lacks sufficient sensitivity for the analysis of plasma from females. It is, however, quite useful in the determination of testosterone in plasma from males to a concentration of $0.1 \mu g$ of testosterone per 100 ml blood or in conditions in the female where such levels are obtained.

The method incorporates extraction of the steroid from plasma with ether. After preliminary washing and elimination of extraneous lipids, testosterone is separated from a number of other steroids by a thin-layer chromatogram (Table 15-10). After elution, final separation and quantitation are achieved by gas–liquid chromatography using a flame ionization detector. Relative retentions of some of the compounds conceivably present in the extracts are shown in Table 15-11.

Specificity of the method is based on multiple thin-layer chromatography, GLC before and after formation of two different derivatives, and addition of authentic steroid to an extract, showing only a single peak. A recent re-evaluation of the accuracy of the procedure using tritiated testosterone showed that approximately 70% of added steroid was recoverable.

In a series of extracts containing about $1 \mu g$ per 100 ml of plasma, simultaneous analyses show an error of $\pm 4.7\%$. A five- to tenfold increase in the sensitivity of this procedure may be anticipated with the present availability of more sensitive detectors and better electrometers. Whether this will allow the routine determination of testosterone from female plasma with such a simple procedure remains to be seen.

TABLE 15-10. R_f Values of Various
Steroids *

Steroid	R_f
Cortisol	0.43
Pregnanediol	0.49
Testosterone	0.59
11-Ketoetiocholanolone	0.62
Corticosterone	0.64
Etiocholanolone	0.66
Pregnanetriol	0.67
Androsterone	0.68
17α-OH-Progesterone	0.69
Dehydroepiandrosterone	0.71
Pregnenolone	0.71
Δ^4-Androstenedione	0.84

*Reproduced by permission from R. Guerra-Garcia,
S. C. Chattoraj, L. J. Gabrilove, and H. H. Wotiz,
Steroids 2:605-616 (1963), Table 2. The compounds
were chromatographed on silica gel (0.25 mm thick).
Development was with benzene: methanol (85:15 v/v)
for 85 min.

Of the various ionization detectors attached to gas chromatographic
instruments, the electron capture detector is surely the most sensitive.
Brownie et al. [32] have utilized this detection principle in a more
involved but very elegant and sensitive method which allows the
accurate determination of testosterone down to the range of concentrations of this hormone in female plasma.

The method includes the addition of tritiated testosterone of high
specific activity followed by extraction of the total lipids from the
plasma. After several washings the neutral extract is chromatographed
on a thin-layer plate and extracts from it are chloroacetylated. The
steroid derivative is then subjected to further thin-layer chromatography, a small portion is counted to ascertain recovery, and part
of the remaining material is gas chromatographed on a 1% XE-60
column. The addition of cholesterol monochloroacetate permits a
more accurate quantitative determination.

In a considerable number of assays, the mean recovery of steroid
was found to be 56.5%. A series of 10 analyses on 5 ml each of male

plasma gave a mean testosterone concentration of $0.021\,\mu g \pm 0.006$. The sensitivity of the method, based on the fact that the minimum visible concentration of steroid in the electron capture cell is $0.001\,\mu g$, is $0.01\ \mu g$ per 100 ml of plasma. An error of $\pm 25\%$ must be anticipated at this low level. The specificity of any method measuring such very low concentrations of steroid is always difficult to ascertain. Table 15-12 shows the R_f values in the two thin-layer systems as well as the relative retentions of a number of steroid chloroacetates on XE-60. Thus the specificity is primarily based on the chromatographic behavior and the exclusion of a number of other steroidal substances which might conceivably be present in the extract. Figure 15-8 shows some typical gas chromatograms obtained by the authors for standards as well as plasma extracts.

EXPERIMENTAL

Method TP-I (Guerra-Garcia et al.)

Materials and Methods. Blood is collected individually in heparinized bottles and immediately centrifuged. The plasma is transferred to a separatory funnel and $0.00412\,\mu g$ of $1,2^{-3}H$-testosterone (s.a. $153\,\mu C/\mu g$) is added. The plasma is extracted four times with

TABLE 15-11.   Relative Retention of
Free Steroids*

Steroid	Relative retention
Etiocholanolone	0.46
Dehydroepiandrosterone	0.47
Androsterone	0.48
11-Ketoetiocholanolone	0.52
17 α-OH-Progesterone	0.60
Δ^4-Androstenedione	0.62
Testosterone	0.66
Pregnenolone	0.69

*Reproduced by permission from R. Guerra-Garcia, S.C. Chattoraj, L.J. Gabrilove, and H.H. Wotiz, Steroids 2:605–616 (1963), Table 1. Cholestane: t_r 10.5. Column: 6 ft by $\frac{1}{16}$ in. ID, 3% SE-30, 250°C, 20 psi N .

TABLE 15-12.   Thin-Layer and Gas Chromatography of Steroids[*]

Steroid	Thin-layer chromatography		Gas chromatography on 1% XE-60
	R_T[†] in ethyl acetate : cyclohexane (1 : 1)	After chloroacetylation R_{TCA}[‡] in ethyl acetate: benzene (1 : 4)	Retention time of chloroacetates (min)
Adrenosterone	0.12		10
Estriol	0.20		
Cortisol	0.20		
11-Dehydrocorticosterone	0.20		
Corticosterone	0.20	0.23	
Cortisone	0.20		
Compound S	0.54	0.36	
Aldosterone	0.74	0.88	
11-Ketoprogesterone	0.77	0.47	
Pregnanediol	0.85	1.85	
19-Nortestosterone	0.85	0.96	12
Epitestosterone	0.98	1.04	12
Testosterone	1.00	1.00	14
20 α-Hydroxypregn-4-en-3-one	1.15	1.04	24
20β -Hydroxypregn-4-en-3-one	1.21	1.17	20
17 α-Hydroxyprogesterone	1.23	0.24	
Etiocholanolone	1.26	1.41	8
Androst-4-ene-3, 17-dione	1.31	0.51	
Androsterone	1.38	1.50	
Dehydroepiandrosterone	1.42	1.35	
3β -Hydroxypregn-5-en-20-one	1.43	1.43	
Estradiol-17β	1.54		
Progesterone	1.65		
Estrone	1.85		
Cholesterol	2.00		11

[*]Reproduced by permission from A. C. Brownie, H. J. VanderMolen, E. E. Nishizawa, and K. B. Eik-Nes, J. Clin. Endocrinol. Metab. 24:1091 (1964), Table 5.
[†]Running rate relative to testosterone.
[‡]Running rate relative to testosterone chloroacetate.

equal volumes of ether, the combined ether extracts are washed three times with $\frac{1}{20}$ volume of 1 N NaOH, followed by equal volumes of 2% $NaHCO_3$ and distilled water. After evaporation of the ether extract at 30°C *in vacuo*, the residue is transferred to a 100 ml centrifuge bottle with 50 ml of 70% methanol and the mixture is stored overnight in a freezer. After centrifugation, the supernatant is transferred to a separatory funnel. The aqueous methanolic extract is then partitioned twice against $\frac{1}{2}$ volume of petroleum ether and the methanolic layer evaporated to about $\frac{1}{3}$ of its volume. The remaining aqueous phase is extracted three times with two volumes of benzene. After evaporation of the combined benzene extracts the residue is transferred to a 15 ml centrifuge tube with acetone and this latter solvent evaporated under a stream of nitrogen.

Thin-Layer Chromatography (TLC). The plasma extracts, dissolved in 100 μl of acetone, as well as testosterone and etiocholanolone standards (10μg each), are applied to a 20 × 20 cm glass plate coated with Silica Gel G (0.25 mm thickness). The samples are applied 3 cm from the edge of the plate and the chromatogram is developed in benzene:methanol (85:15 v/v) for 85 min using the ascending technique. After development, the plate is dried at room temperature.

The section of the chromatoplates containing the extract is carefully covered up and the portion containing the standards is cautiously sprayed with a solution of concentrated sulfuric acid in absolute ethanol (1:1 v/v) and heated in an oven at 90°C. Testosterone produces a deep green color and etiocholanolone appears brown; this staining technique can detect 1 μg of testosterone. A 1.5-cm-wide zone of the

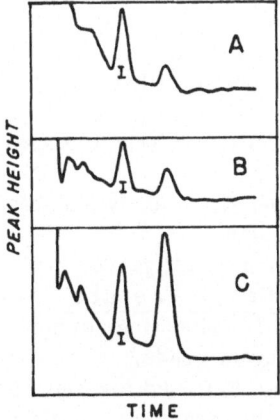

Figure 15-8. A. Cholesterol chloroacetate (0.24 μg) and testosterone chloroacetate (0.0043μg) isolated after processing 10 ml female plasma containing 2500 cpm testosterone-^{3}H. B. Cholesterol chloroacetate (0.17 μg) and testosterone chloroacetate (0.0037 μg) isolated after processing 1 ml male plasma containing 2500 cpm testosterone-^{3}H. C. Cholesterol chloroacetate (0.27μg) and testosterone chloroacetate (0.024 μg) isolated after processing 5 ml male plasma containing 2500 cpm testosterone-^{3}H. I = Cholesterol chloroacetate. II = Testosterone chloroacetate. [A. C. Brownie, H. J. VanderMolen, E. E. Nishizawa, and K. B. Eik-Nes, J. Clin. Endocrinol. Metab. 24:1091 (1964)].

plasma extract corresponding to the R_f of testosterone is removed and thoroughly eluted with acetone. The eluate is dried in a stream of nitrogen, dissolved in 0.1 ml of acetone and a one-tenth portion removed for counting in a Packard Tri-Carb Liquid scintillation spectrophotometer employing 18 ml of a PPO–POPOP system. The remainder of the plasma extract is dried in a small conical tube.

Gas Chromatography (GLC). The dried residue from thin-layer chromatography is dissolved in 50 μl of acetone. A 5 μl portion is injected directly onto a 6-ft by $\frac{1}{8}$-in.-OD stainless steel or 4-mm-ID glass column filled with 3% SE-30 elastomer deposited by evaporation on diatomaceous earth (Diatoport S, 80-100 mesh). The column is maintained at $\sim$ 250°C, with the vaporizer at 250°C. A flow of 30 ml/min of carrier gas is maintained while the detector is operated with hydrogen and air pressure to give optimal response. Quantification of the resultant peaks is carried out by comparison of the peak height of the testosterone peak of the unknown to a standard specimen of similar concentration, after having demonstrated a linear response of the detector in the specific range of concentration. Further correction is made for the removal of $\frac{1}{10}$ aliquot for radioactivity measurement and for the recovery of tritiated testosterone determined thereby.

Method TP-II (Brownie et al.)

Materials: Solvents and Reagents

Ether. Reagent grade, distilled through a 60-cm fractionating column.

Methanol. Anhydrous, analytical reagent, distilled from 2,4-dinitrophenylhydrazine.

Benzene. Thiophene–free, analytical reagent, distilled through a 60-cm fractionating column.

Toluene. Analytical reagent grade.

Chloroform and Ethyl Acetate. Analytical reagent grade, distilled through a 60-cm fractionating column.

Cyclohexane. Reagent grade.

Tetrahydrofuran. Refluxed for 3 hr with potassium hydroxide pellets and distilled. It is then distilled off sodium and stored over sodium in a dark brown bottle.

Pyridine. Refluxed over barium oxide for 4–6 hr and distilled through a fractionating column. The fraction boiling at 115°C is collected and stored over calcium chloride in a desiccator.

Monochloroacetic Anhydride. Dried and stored in a desiccator.

2,5-Diphenyloxazole(PPO) and 1,4-bis-[2-(5-Phenyloxazolyl)]-benzene (POPOP). Scintillation grade, used without further purification.

Silica Gel G. (Merck, according to Stahl, for thin-layer chromatography.) Washed with boiling methanol after the addition of a phosphor (DuPont luminescent chemical, index 609), 30 mg per 100 g. The washed silica gel is dried at 100°C for 24 hr before use.

Testosterone-1,2-^{3}H. Specific activity ∼ 150 μC/μg. A solution (ca. 2500 cpm/ml) of testosterone-1,2-^{3}H in methanol is stored at 5°C.

Testosterone. Recrystallized from hexane with a few drops of methanol to constant melting point.

Testosterone Chloroacetate. One gram chloroacetic anhydride and 0.2 ml pyridine are added to 100 mg testosterone in 5 ml tetrahydrofuran. After the solution has stood in the dark for 18 hr, 5 ml of water is added and the solution is extracted with 5 ml ethyl ether 3 times. The ether extract is washed once with 5 ml of 6 N hydrochloric acid and twice with 5 ml of water. The washed ether extract is filtered through anhydrous sodium sulfate and taken to dryness. The residue is recrystallized several times from aqueous acetone to give crystals with melting points of 124°–125°C (corrected).

Cholesterol Chloroacetate. Prepared by the method described for testosterone chloroacetate. From the ether extract a white crystalline residue is obtained which is recrystallized several times from aqueous acetone to give crystals melting at 159–160°C (corrected).

Thin-Layer Chromatography. This process is carried out on silica gel plates. The plates are prepared with a spreader, giving layers of approximately 0.30 mm thickness; about 3 g silica gel and 9 ml water are needed to prepare each 20 × 20 cm plate. The plates are dried at 100°C for at least 90 min before use. Extracts are chromatographed on 2 cm lanes, separated from each other by a 1 cm lane and from standards by a 1.5 cm lane. This arrangement allows 4 extracts and 2 standards to be run on each plate. Samples are applied to thin-layer plates with chloroform–methanol (1:1 v/v) using successively 4, 3, and 2 drops of this solution. Chromatograms are developed in an ascending manner. Steroid standards on the chromatograms are detected using a Haines fluorescent scanner.

Assay of Radioactivity. A liquid scintillation spectrophotometer set to give approximately 25% counting efficiency is used to assay ^{3}H. Samples for assay are dried and dissolved in 10 ml of scintillation fluid (4 g PPO and 40 mg POPOP in 1 liter toluene) in 20 ml glass vials.

Estimation of Testosterone in Human Blood. Preparation of Blood Plasma. Whole blood is withdrawn into a bottle containing heparin and the plasma is obtained by centrifugation. A solution of testosterone-1,2-^{3}H (2500 cpm) is taken to dryness in a 50 ml glass tube and 10 ml plasma is added (5 ml plasma plus 5 ml 0.9% sodium chloride in the case of male plasma).

Extraction. 20% Sodium hydroxide, 0.25 ml, is added to the plasma, which is extracted 3 times with 20 ml diethyl ether. The pooled ether extracts are washed twice with 5 ml distilled water and concentrated under nitrogen at 40°C. The extract is then transferred to a 15 ml conical centrifuge tube and taken to dryness, the extract concentrated in the tip of the bottle.

Thin-Layer Chromatography of Plasma Extract. Thin-layer chromatography is carried out on silica gel plates as described. An ascending chromatogram is run in the solvent cyclohexane: ethyl acetate (1:1 v/v). Standards of testosterone (0.5 μg) are chromatographed on separate lanes. At this stage of purification testosterone in plasma extracts runs slower than testosterone standards; consequently, the area of silica gel scraped off from the plasma extract lanes extends from just beyond the testosterone standard to a point 1 cm nearer the origin than the standards. Three drops of water are added to the silica gel and this is then extracted with 1 ml methanol 3 times. The combined extracts are taken to dryness in a 15 by 150 mm test tube.

Monochloroacetylation. After the extract is dried in a vacuum desiccator, 0.5 ml of a solution of monochloroacetic anhydride in tetrahydrofuran (100 mg per 10 ml) and 0.1 ml pyridine is added. The reaction is carried out overnight in a darkened desiccator. One milliliter of water is added to stop the reaction and the solution is extracted three times with 1 ml of ethyl acetate. The combined ethyl acetate extracts are washed once with 1 ml of 6 N hydrochloric acid and twice with 1 ml of distilled water and taken to dryness in a 15 ml centrifuge tube.

Thin-layer Chromatography of Testosterone Monochloroacetate. The extract from the chloroacetylation is chromatographed on silica gel plates using the solvent system benzene: ethyl acetate (4:1 v/v). Standards of authentic testosterone chloroacetate are run and the corresponding areas from the extracts scraped off. After the addition of 4 drops of water, the silica gel is extracted three times with 1 ml of benzene. High-speed centrifugation is used to obtain a benzene layer free from silica gel. The benzene extracts are transferred to 2.0 ml conical microcentrifuge tubes and dried down after each extraction.

Addition of Internal Standard and Sampling for ^{3}H Counting. The dried extracts in the microcentrifuge tubes are dissolved in 1 ml of methanol containing 0.4 μg/ml cholesterol monochloroacetate.   One-tenth aliquot of this mixture is taken for ^{3}H assay and the remainder is evaporated to dryness.

Gas Chromatography. The dried extract is dissolved in 15 μl toluene, and as much as possible (usually 8–10 μl) of this solution is injected into a 3-ft gas chromatographic column.

The stationary phase for this column is prepared as described by DePaoli et al. [18], using 1 g of XE-60 dissolved in 100 ml toluene for 25 g purified deactivated Gas Chrom P.

The column temperature is kept at 210°C with the detector at 220°C and the flash heater at 250°C. High-purity nitrogen is used as carrier gas and led through a tube filled with molecular sieve (type 13× from Linde); the gas pressure is kept at 40 psi. The electron capture detector is operated under conditions of maximal sensitivity and adequate linear range.* Quantitation is carried out by area measurement of the peak according to the following formula, which allows for the various losses incurred, and by comparison to the added internal standard:

$$S(\mu g) \ = \frac{X_s}{10X_p} \times \frac{C_s}{C_x} \times \frac{T_x}{T_s} \times \frac{288.4}{364.5} \times 0.01$$

where X_s is cpm of testosterone-1,2-^{3}H initially added to the plasma, X_p is cpm of ^{3}H in the aliquot prior to gas chromatography, C_s is area (cm^2) of 0.4 μg cholesterol monochloroacetate, C_x is area (cm^2) of cholesterol monochloroacetate in sample, T_s is area (cm^2) of 0.01 μg testosterone monochloroacetate, T_x is area (cm^2) of testosterone mono-chloroacetate in sample, 288.4 is molecular weight of testosterone, and 364.5 is molecular weight of testosterone monochloroacetate.

Further correction for the mass measurement of a water blank containing an equal amount of ^{3}H-testosterone and carried through the full procedure is also necessary. However, it should be noted that this latter correction can be eliminated because of the availability of very high specific activity testosterone (150 μC/μg).

REFERENCES

1. H. H. Wotiz, H. M. Lemon, P. Marcus, and K. Savard, J. Clin. Endocrinol. Metab. 17:534 (1957).
2. E. Dingemanse, L. G. Huis int Veld, and B. M. DeLaat, J. Clin. Endocrinol. 6:535 (1946).
3. A. M. Robinson and F. Goulden, Brit. J. Cancer 3:62 (1949).
4. T. K. Lakshmanan and S. Lieberman, Arch. Biochem. Biophys. 53:258 (1954).
5. A. E. Kellie and A. P. Wade, Biochem. J. 66:196 (1957).
6. K. Savard, J. Biol. Chem. 202:457 (1953).
7. B. L. Rubin, R. I. Dorfman, and G. Pincus, J. Biol. Chem. 203:629 (1953).
8. L. T. Gardner, J. Clin. Endocrinol. Metab. 13:941 (1953).
9. C. J. Migeon and J. E. Plager, J. Clin. Endocrinol. Metab. 15:702 (1955).
10. M. Sparagana, E. H. Keutmann, and W. B. Mason, Anal. Chem. 35:1231–1238 (1963).

*See Chapter 4.

11. E. O. A. Haahti, W. J. A. VandenHeuvel, and E. C. Horning, Anal. Biochem. 2:182–187 (1961).
12. I. S. Hartman and H. H. Wotiz, Steroids 1:33–38 (1963).
13. W. J. A. VandenHeuvel, B. G. Creech, and E. C. Horning, Anal. Biochem. 4:191–197 (1962).
13a. B. G. Creech, J. Gas Chromatog. 2:194–195 (1964).
14. J. T. France, N. L. McNiven, and R. I. Dorfman, Acta Endocrinol. 90:71–80 (1964).
15. J. T. France, R. Rivera, N. L. McNiven, and R. I. Dorfman, Steroids 5:687 (1965).
16. M. A. Kirschner and M. B. Lipsett, J. Clin. Endocrinol. Metab. 23:255–260 (1963).
17. P. Vestergaard and B. Claussen, Acta Endocrinol. Suppl. 64:35 (1962).
18. J. C. DePaoli, E. E. Nishizawa, and K. B. Eik-Nes, J. Clin. Endocrinol. Metab. 23:81 (1963).
19. B. S. Thomas and R. D. Bulbrook, J. Chromatog. 14:28 (1964).
20. L. Segal, B. Segal, and W. R. Nes, J. Biol. Chem. 235:3108 (1960).
21. I. E. Bush, The Chromatography of Steroids (Pergamon Press, New York, 1961), p. 337.
22. R. I. Dorfman, Methods in Hormone Research, Vol I. (Academic Press, New York, 1962), p. 53.
23. W. Futterweit, N. L. McNiven, L. Marcus, C. Lantos, M. Drosdowsky, and R. I. Dorfman, Steroids 1:628 (1963).
24. H. Ibayashi, M. Nakamura, S. Murakawa, T. Uchikawa, T. Tanioka, and K. Nakao, Steroids 3:559 (1964).
25. R. V. Brooks and G. Giuliani, Steroids 4:101 (1964).
26. A. M. Camacho and C. J. Migeon, J. Clin. Endocrinol. Metab. 23:301 (1963).
27. D. H. Sandberg, N. Ahmad, W. W. Cleveland, and K. Savard, Steroids 4:557 (1964).
28. M. Sparagana, Steroids 5:773 (1965).
29. M. Finkelstein, E. Forchielli, and R. I. Dorfman, J. Clin. Endocrinol. Metab. 21:98 (1961).
30. K. J. Ryan, J. Biol. Chem. 234:268 (1959).
31. R. Guerra-Garcia, S. C. Chattoraj, J. L. Gabrilove, and H. H. Wotiz, Steroids 2:605 (1963).
32. A. C. Brownie, H. J. VanderMolen, E. E. Nishizawa, and K. B. Eik-Nes, J. Clin. Endocrinol. Metab. 24:1091 (1964).

Chapter 16

Determination of Urinary Pregnanetriol

BIOCHEMISTRY AND EVALUATION OF METHODS

During the metabolism of 17α-hydroxy progesterone reduction of

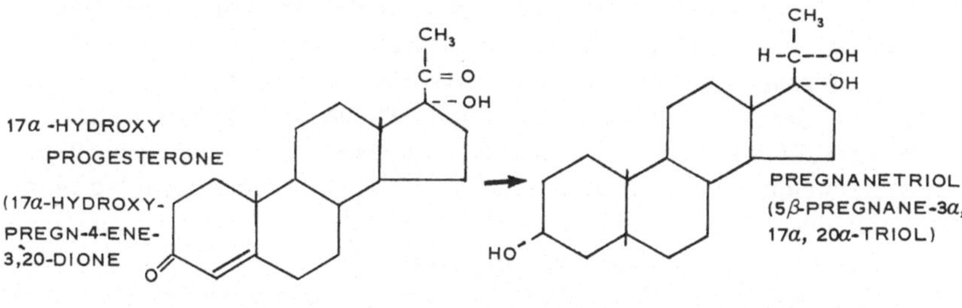

17α-HYDROXY
PROGESTERONE

(17α-HYDROXY-
PREGN-4-ENE-
3,20-DIONE

PREGNANETRIOL
(5β-PREGNANE-3α,
17α, 20α-TRIOL)

the double bond in ring A and the ketone groups at position 3 and 20 occurs, with the formation of pregnanetriol. In turn, 17α-hydroxy progesterone is derived by hydroxylation at the 17α-position of the primary hormone progesterone. As has been discussed earlier, the latter compound has been shown to be produced by the corpus luteum of the ovary, in rather large quantities by the human placenta, and by the adrenal gland. In general, there appears to be relatively little accumulation of pregnanetriol in normal circumstances.

In the adrenal gland this compound is further hydroxylated by a series of adrenocortical enzymes at positions 11 and 21, with the final formation of cortisol. Under certain circumstances, generally associated with adrenal hyperplasia, an inability of the gland to produce sufficient enzyme for C-21 hydroxylation occurs. Since Hayano [1] has postulated that hydroxylation at C-21 is necessary before C-11 hydroxylation, an overabundance of 17α-hydroxy progesterone would develop in the adrenal gland. As a consequence increased titers of urinary pregnanetriol may be expected. The association of this adrenal syndrome with the metabolic defect, deduced from the elevated pregnanetriol excretion, has been described in a series of elegant papers by Bongiovanni and his co-workers [2]. Consequently, several methods for the determination of pregnanetriol were developed over the last few years. Among them one might mention the procedure of Stern [3], who after enzyme hydrolysis of the conjugate and preliminary partition separated the pregnanetriol from pregnanediol and other contaminants on an alumina column. The proper fraction was then assayed as the sulfuric acid chromogen. Bongiovanni and Eberlein [4] used a rather similar approach, also employing an alumina column for the separation, but carrying out their elution with different solvent mixtures. Cox [5], on the other hand, developed a procedure which was based on the cleavage of the side chain which, though less specific, did allow for the more rapid determination of acetaldehyde following periodic acid cleavage at C-17. Norymberski [6] developed a procedure for the measurement of the 21-desoxy-17,20-dihydroxy steroids, using sodium bismuthate to remove the side chain. First attempts to measure pregnanetriol by gas chromatography were reported by Rosenfeld et al. [7], who verified that pregnanetriol may be gas chromatographed intact. The method did not separate pregnanetriol (PT) from $3\alpha,17\alpha$-dihydroxy-pregnan-20-one and Δ^5-pregnene-3β, $17,20\alpha$-triol (Δ^5-PT). Consequently, somewhat time-consuming paper chromatographic separation preliminary to gas chromatographic assay was included in the procedure. A recent modification of this procedure is discussed here and described in detail below.

In 26 determinations, where urinary PT analyses by GLC have been compared with the acetaldehydogenic procedure of Cox [5], values ranged from 0.75 to 25.2 mg per day in subjects with a variety of clinical conditions and averaged about 10% lower than the values obtained from the method of Cox. In 21 analyses for Δ^5-PT by both procedures, values ranged from 0.27 to 33.5 mg per day by GLC and averaged 5% lower than those obtained from the acetaldehydogenic procedure. Where urinary PT and Δ^5-PT values are in the high range, smaller urine aliquots may be processed or the TMSi solution may be diluted for injection onto the column. Daily excretion of PT in subjects with normal adrenocortical function varied from 0.13 to 1.0 mg; the same subjects excreted from 0.10 to 0.91 mg of Δ^5-PT. It should be pointed out that some of these data were accumulated by analysis of underivatized triols on SE-30, as well as by GLC analyses of the TMSi derivatives, a modification of the earlier procedure.

The Δ^5-pregnenetriols which contain the 16α-hydroxy group may be readily identified and measured by the combination of paper chromatography and GLC. Happily, the compounds which migrate together have significant differences in retention times. This can be seen in Table 16-1.

In 1964, Kirschner and Lipsett [8] described a procedure already discussed under pregnanediol and 17-ketosteroid analysis, which simultaneously allows the quantitative determination of pregnanetriol.

TABLE 16-1. Chromatography of Some Pregnenetriols

Substance	Distance from origin in paper chromatography (cm)	Relative Retention	
		SE-30*	QF-1†
Δ^5-Pregnene-3β, 16α, 20β-triol	20−30	1.63	1.32
Δ^5-Pregnene-3α, 16α, 20β-triol	20−30	1.22	1.00
Δ^5-Pregnene-3β, 17, 20α-triol	65−73	1.55	1.90
Δ^5-Pregnene-3α, 16α, 20α-triol	73−93	1.13	0.96
Pregnane-3α, 17, 20α-triol	93−98	1.30	1.64

*Cholestane = 8.4 min.
†Cholestane = 2.8 min.

As has been already mentioned under the previous sections of this particular method of hormone determination, the specificity for analysis of individual components is primarily derived through their separation on thin-layer chromatograms and the exactness of the retention time during gas chromatography. Furthermore, the addition of authentic steroids to urinary extracts resulted in symmetrical peak augmentation with no appearance of new peaks. The precision of the method as it applies to the analysis of pregnanetriol is very good, with a coefficient of variance of 3%.

Recovery studies to determine the accuracy of the procedure showed that 104% of added pregnanetriol was recovered with a variance of ± 5%.   Sensitivity appears to be 50 μg per 24 hr of pregnanetriol excretion.   When the procedure described below was used, seven male subjects had a mean pregnanetriol excretion of 0.89 mg per 24 hr with a range of 0.36 to 1.83. In nine female subjects a mean value of 0.77 mg per 24 hr was obtained with a spread from 0.21 to 2.10.

EXPERIMENTAL

Method PT-I (Rosenfeld)

Hydrolysis and Preliminary Fractionation.   The urine is hydrolyzed, extracted, and the nonketonic extract is obtained according to the methods of Kappas and Gallagher [9] and Dobriner [10]. The urine is first adjusted to pH 5 with sulfuric acid; then 10 vol. % of sodium acetate buffer at pH 5 and 300,000 units per liter of beef liver β-glucuronidase are added. The solution is incubated at 37° for 120 hr, then continuously extracted with diethyl ether for 48 hr. The ether solution is washed with 2 N sodium hydroxide, followed by three washes with 5% sodium chloride solution. This affords the enzyme-hydrolyzed neutral fraction $\underline{A}$.

The spent urine and washes from the above are combined and the solution is brought to pH 1 with sulfuric acid. This is continuously extracted with diethyl ether for 72 hr. A neutral fraction $\underline{B}$ is obtained from the ether solution as described above. $\underline{A}$ contains steroids originally conjugated with glucuronic acid, while $\underline{B}$ contains those present as sulfate esters before hydrolysis.

The nonketonic fractions from $\underline{A}$ and $\underline{B}$ are obtained according to a modification of the Girard method [3, 4]. For 100 mg of neutral fraction, 2 ml of ethanol, 100 mg of Girard's reagent T, and 0.1 ml of glacial acetic acid are mixed and refluxed for 2 hr. Ice is added, followed by 0.75 ml of 10% sodium hydroxide. The cold suspension is extracted with two 150 ml portions of ether. The ether layers are washed twice with 25 ml portions of water. Concentration of the combined ether solutions affords the nonketonic extract.

A part of the extract, corresponding to about 25% of a day's urine, is applied in a line to Whatman No. 1 paper (18 × 118 cm) and, after equilibration, the chromatogram is developed in the system 2,2,4-trimethylpentane: toluene: methanol: water (3:5:4:1) (methanol: water being the stationary phase) for 48 hr at 24°. The reference compounds are visualized with phosphomolybdic acid and the corresponding areas of the extract are cut from the paper and eluted with warm methanol. The triols will have moved the following distances from the origin: PT, 93–98 cm; Δ^5-PT, 65–73 cm; Δ^5-PT-3 a, 16 a, 20 a, between the former; Δ^5-PT-3 β, 16 a, 20 β and Δ^5-PT-3 a, 16 a, 20 β, 20–30 cm.

The paper chromatographed fractions of the extracts are concentrated *in vacuo* to a small volume and filtered through a Celite pad to remove debris. They are then transferred to a 1 ml volumetric tube and dried thoroughly before conversion to the trimethylsilyl ether derivatives and gas chromatography.

Alternative Procedure Using TLC. For PT and Δ^5-PT, the triols most commonly measured, an alternative procedure may be used where the extract is applied in a narrow band to a 20 × 20 cm Silica Gel G coated glass plate and the chromatogram is developed for 20 min in the system ethyl acetate: cyclohexane (7:3). The zones containing the proper steroids are located with sulfuric acid:ethanol (1:1). After the areas of the extract corresponding to the authentic standards of the edge of the TLC plate are scraped off, the steroids are eluted with acetone, and after filtration through a sintered glass funnel into a 3 ml centrifuge tube, the solvent is evaporated.

Derivative Preparation. To each dried fraction is added 100 μg cholestanol (internal standard), 0.8 ml pyridine, 80 μl hexamethyl-

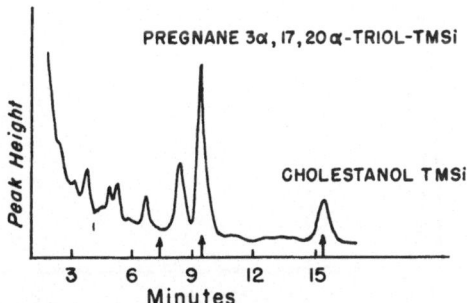

Figure 16-1. GLC tracing of paper chromatograph extract for PT (TMSi derivative) on SE-30; $T = 226°C$.

disilazane, and 16 μl trimethylchlorosilane. The tubes are capped with Teflon stoppers and allowed to remain overnight at room temperature. The reagents are removed at 60° under a slow stream of nitrogen and are then mixed with 0.2 ml of chloroform for injection.

Gas Chromatography. From 3 to 8 μl of the derivatized material is injected with a Hamilton syringe (10 μl capacity) onto an SE-30 column (3 wt.% on 100-140 mesh Gas Chrom P; dimensions, 1.8m × 5 mm) maintained at 226°; carrier gas pressure, 30 psi. Authentic samples of PT, Δ^5-PT, epimeric Δ^5-PT's, and cholestanol are converted to the trimethylsilyl ethers for standards. Reference curves, allowing plotting of weight vs. area under the peaks, are constructed to which the appropriate peak areas in the samples are compared (Figures 16-1 and 16-2).

Alternative GLC Procedure. More recently, this procedure has been modified for analyses with a QF-1 column. The dried extract in a 3 ml centrifuge tube is treated with the same reagents as in the previous section, but without the internal standard. To the dried derivatives is added 0.2 ml of chloroform which contains 200 μg of

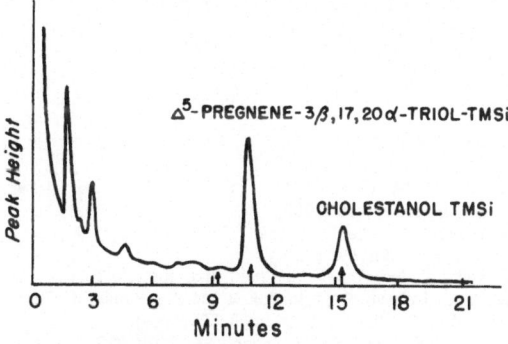

Figure 16-2. GLC tracing of paper chromatograph extract for Δ^5 PT (TMSi derivative). Conditions are the same as in Figure 16-1.

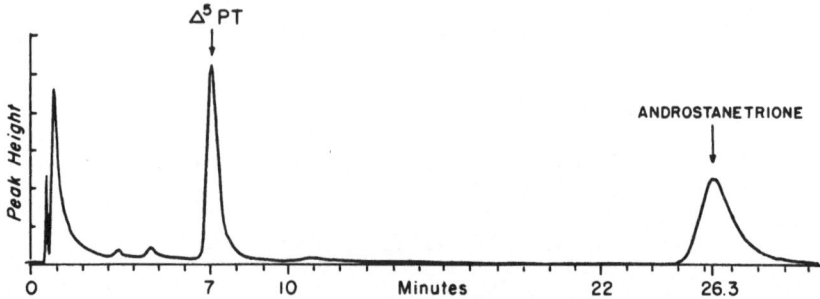

Figure 16-3. GLC tracing of paper chromatograph extract for Δ^5 PT (TMSi derivative) on QF-1. Peak at 26.3 ft = androstane-3,11,17-trione; T = 225°.

androstane-3,11,17-trione. The tube is agitated, then centrifuged for 5 min. Portions are injected into the gas chromatograph, care being taken to avoid the packed precipitate.

With the QF-1 column (3 wt.% on 100-140 mesh Gas Chrom P), the temperature used is usually 225°, all other conditions remaining the same as for the SE-30 column (Figure 16-3).

Method PT-II (Kirschner and Lipsett)

The hydrolysis, extraction, preliminary separation of pregnanetriol by solvent partition, and thin-layer chromatographic separation have been described in detail in Chapter 15. The reader should follow the instructions given for this method in that chapter (p. 177). The third zone (seen earlier in Figure 15-1) is eluted after location of the appropriate marker exactly as described previously. The area corresponding to pregnanetriol may then be analyzed directly by injection onto an XE-60 column under the conditions described or, in the event that low concentrations are anticipated, it would be preferable to convert the triol to its trimethylsilyl ether derivative, followed by GLC analysis on an SE-30 column at 235°. Quantitation by area determination is carried out as before.

REFERENCES

1. M. Hayano, N. Saba, R. I. Dorfman, and O. Hechter, Recent Progr. Hormone Res. 12:79 (1956).
2. A. M. Bongiovanni and G. W. Clayton, Bull. Johns Hopkins Hosp. 94:180 (1954).
3. M. I. Stern, J. Endocrinol. 16:180 (1957).
4. A. M. Bongiovanni and W. R. Eberlein, Anal. Chem. 30:388 (1958).
5. R. I. Cox, J. Biol. Chem. 234:1693 (1959).
6. C. J. W. Brooks and J. R. Norymberski, Chem. Ind. 804 (1952).
7. R. A. Rosenfeld, M. C. Lebeau, R. D. Jandorek, and T. Salumaa, J. Chromatog. 8:355 (1962).
8. M. A. Kirschner and M. B. Lipsett, Steroids 3:277 (1964).
9. A. Kappas and T. F. Gallagher, J. Clin. Invest. 34:1566 (1955).
10. K. Dobriner, J. Clin. Invest. 32:940 (1953).

Chapter 17

Analysis of Urinary Adrenocortical Steroids

BIOCHEMISTRY AND EVALUATION OF METHODS

Cumulative evidence from the past two decades suggests that the adrenal cortex is more than a single gland. Indeed, it is possible to differentiate the area responsible for the biogenesis of androgenic material from the zone responsible for the biosynthesis of aldosterone and this from yet another zone responsible for the formation of cortisol. The assessment of the functional ability of each of these zones would, therefore, depend on the adequate analyses of the metabolic products derived from the specific steroid associated with each segment. We shall concern ourselves primarily with a discussion of methodology relating to the metabolic excretion products derived from cortisol, which is conceded to be the primary glucocorticoid secreted by the human adrenal cortex. Through the mediation of peripheral and hepatic metabolism a series of reactions occur, modifying the molecule of cortisol as outlined in Table 17-1. The side-chain degradation leading to androgenic or C-19 steroids has already been discussed previously and need not be gone into any further. Earlier studies on the metabolism of cortisol or cortisone indicated that one of the prime urinary excretion products was the compound tetrahydrocortisone. The excellent study of administered cortisol-4-^{14}C by Fukushima et al. [1] in 1955 further showed that a series of extensive reductions in the side chain also occur, yielding significantly large concentrations of epimeric cortolones and cortols in the urine. The reader of the literature on this particular subject soon becomes aware of the fact that there are a fairly extensive number of urinary metabolites of cortisol which must be reckoned with in trying to determine the most advantageous analytical method, particularly when the assay is to be used for the elucidation of clinical aberrations. The question, here more than with any other steroid assay, which confronts the investigator is whether to analyze large fractions of the known urinary metabolites as one unit (cf. 17-ketogenic steroids, 17-hydroxy corticosteroids, etc.) or to analyze one or several of the individual components, any one of which might be of particular

207

TABLE 17-1

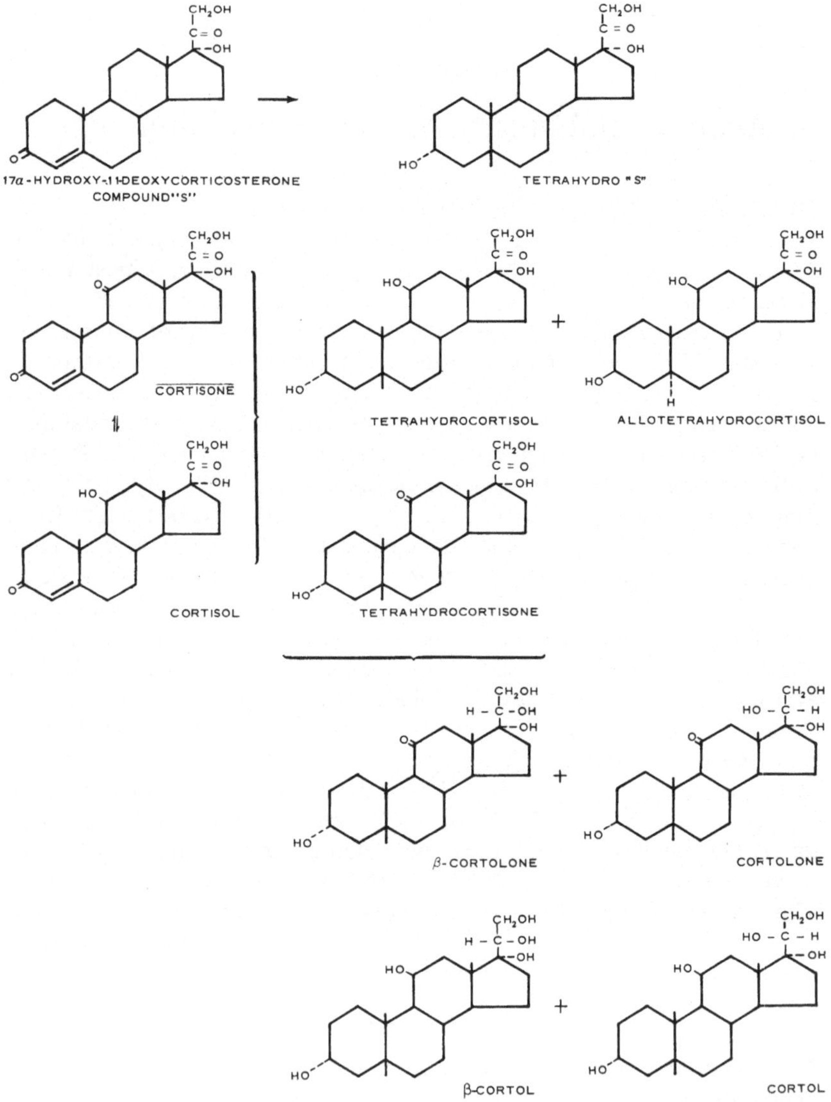

17α-HYDROXY-11-DEOXYCORTICOSTERONE COMPOUND "S"

TETRAHYDRO "S"

CORTISONE

TETRAHYDROCORTISOL

ALLOTETRAHYDROCORTISOL

CORTISOL

TETRAHYDROCORTISONE

β-CORTOLONE

CORTOLONE

β-CORTOL

CORTOL

significance in the study of a disease associated excretion pattern. The important methods of screening for glucocorticoid activity developed by Reddy et al. [2] and Norymberski [3], to mention but a few among many, characterized the type of procedures measuring a group of compounds rather than individual ones, and for speed and significance probably are not matched by comparable gas chromatographic assays. On the other hand, the isolation of individual compounds or groups of compounds, which up to now were generally achieved by laborious multiple paper chromatographic procedures, can in some instances be matched by gas chromatographic methods already published.

In 1960, VandenHeuvel and Horning [4] reported the first attempts to separate corticosteroids by gas chromatography. Although they were able to show that deoxycorticosterone and corticosterone have retention times sufficiently great to suggest maintenance of their structural integrity, the 17a-hydroxylated corticosteroids in all instances produced peaks occurring at a retention time essentially corresponding to the analogous 17-ketosteroids. The difficulty in using this approach for assay of a specific corticosteroid lies in the fact that more than one urinary metabolite may give rise to a single degradation product.

In 1961, Wotiz and Carr [5] showed that through the use of derivatives (acetates) and a sufficiently high concentration of stationary phase, both cortisone and aldosterone could be chromatographed without alteration of the molecule. In 1961, Merits [6], employing periodic acid oxidation of several corticosteroids, obtained the expected etiocholanic acids (with the exception of aldosterone, which forms the gamma lactone) and following esterification of the acids with diazomethane, chromatographed them without any further alteration. Separation of deoxycorticosterone, corticosterone, cortexolone, cortisone, cortisol, and aldosterone was achieved on a neopentylglycol succinate (NGS) column. As yet, however, this procedure seems not to have been applied to urinary extracts. Kittinger [7], using the general approach outlined by Merits, did develop a procedure for the quantitative determination of submicrogram quantities of corticosteroids produced *in vitro* by rat adrenal glands. He was able to determine the following steroids: progesterone, deoxycorticosterone, 11β-hydroxyandrostenedione, dehydrocorticosterone, 11β-hydroxyprogesterone, corticosterone, and 18-hydroxydeoxycorticosterone as well as aldosterone.

In a slightly different vein, Kirschner and Fales [8] reported the analyses of 17a-hydroxy corticosteroids following the formation of the bis-methylenedioxy derivatives. The method has been used for determining the concentration of cortisone in guinea pig urine. While this approach shows considerable promise in its application, it has not yet been used for the analysis of human urine nor are any statistics available regarding accuracy or sensitivity of the method. It is likely that the procedure incorporates a greater specificity than the use of the acetates because of the unique interaction with the dihydroxy acetone.   The reader must approach usage with caution, however, because of the necessity to utilize 12 N HCl to form the derivative.   In turn, this would lead to dehydration of any 11β-hydroxy compound.   It would, however, appear quite interesting to investigate this procedure for its use for the determination of urinary cortisone.

One important point raised by these authors relates to the pyrolysis of cortisone, which has been the basis of several assay procedures which will be mentioned shortly.   It appears that if cortisone is injected into the gas chromatograph, pyrolysis does occur and adrenosterone is produced, but the detector response on equimolar quantities is only $\frac{1}{5}$ that obtained by pyrolysis as compared to pure adrenosterone.   A more recent publication by Gottfried [9] further shows the fallacy of this approach since the latter showed the non-linearity of detector response following pyrolysis.

Despite the fact that evidence has been presented that 17-hydroxylated corticosteroids could be gas chromatographed intact, no reports have as yet appeared in the literature describing methods of analysis of primary adrenocortical hormones as the intact molecule. One of the main reasons that we have been unable to carry through such analyses in our laboratory, at sensitivities significant in urinary assay procedures, has been the considerable sorption of corticosteroid acetates on nonpolar columns such as SE-30.

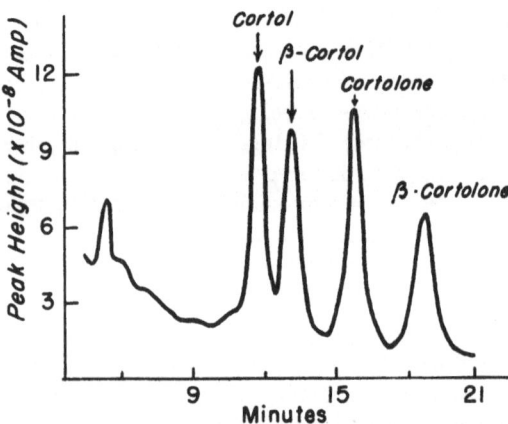

Figure 17-1. Chromatography of TMSi ether derivatives on QF-1; T=220°. (1) Cortol (2) β-cortol (3) cortolone (4) β-cortolone.   Cholestane = 3.0 min. (R. Rosenfeld, Steroids 4:147 (1964).

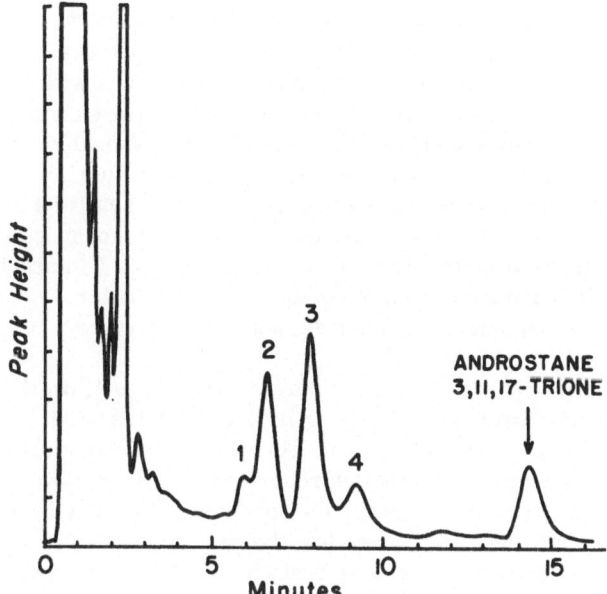

Figure 17-2. GLC tracing of urine extract, processed as described. Numbers refer to Figure 17-1.

A significant breakthrough in this approach was reported in early 1964 by Rosenfeld [10], who developed a method for the gas chromatographic determination of the isomeric cortols and cortolones as the trimethylsilyl ethers. Separations were achieved on QF-1 and SE-30 columns. Rosenfeld carried out an investigation into the quantitative conversion, of the four compounds indicated, to their trimethylsilyl ethers and characterized them. Evidence for the character of the peaks corresponding to the derived compounds (Figure 17-1) was obtained by several means including TLC, GLC, carbon, hydrogen, and silicon analyses, and infrared spectroscopy.

The peaks have also been characterized by their gas chromatographic behavior on SE-30 and by preparation of TMSi ethers where the 11-hydroxy group in the cortols remains free. The relative retentions of the derivatives, as used in this procedure, on QF-1 are as follows: cortol, 2.48; β-cortol, 2.74; cortolone, 3.26; β-cortolone, 3.82; androstane-3,11,17-trione (internal standard), 6.17; cholestane, 1.0 (2.3 min).

Comparison of areas under the appropriate peaks with corresponding standards has given estimations of the excretion rates of cortols and cortolones which range from 0.30 to 2.6 mg per day in 19 subjects with normal adrenocortical function, and 10, 22, and 40 mg per day in 3 patients with adrenocortical hyperfunction. There is virtually no information on the measurement of cortols and cortolones as a group, but comparison of the normal values with analyses of suitably purified paper chromatograms [11] by the enzymatic determination of Hurlock and Talalay [12] shows reasonable agreement. Duplicate analyses of extracts usually agree within 15%, a not unreasonable figure since the values represent a sum of 4 component compounds, one of which, cortol, is difficult to measure. This compound is usually present in smallest amounts (Figure 17-2) and is not well separated from β-cortol, excreted in larger quantities. Recoveries of standard compounds, added to the extract before TLC, are good.

Several investigators have reported methods to allow the assessment of corticosteroids following cleavage of the side chain of the corticosteroids. Luetscher and Gould [13] briefly described the pyrolytic degradation of tetrahydrocortisone, allotetrahydrocortisol, and tetrahydrocortisol. Again, as reported previously by Kirschner and Fales [8], the peak area for the pyrolytically obtained corresponding C-19 steroids is significantly smaller than that for equal concentrations of pure standards. Insufficient information is as yet available on this particular method. The authors report that comparison of analyses by GLC and another method, following preliminary fractionation of the metabolites on liquid–liquid chromatograms and colorimetric assay, showed good agreement. They also report certain problems arising with respect to changes of retention time in the presence of large concentrations of contaminants, possibly due to the inefficiency of pyrolysis which of course must be essentially instantaneous. Indeed, one could visualize a number of problems depending on the type of instrument and injection system available. For instance, it may very well be that with on-column injections the pyrolytic method might not be very useful.

The work reported by Bailey [14] shows two significant advances over the previous method. In the first place, because of the losses occurring during pyrolysis, he decided to oxidize the side chain by use of the sodium bismuthate reaction prior to GLC. Furthermore, in order to get greater reproducibility, less interference from the solvent front, and greater sensitivity, Bailey utilized a system of solid state injection by adsorption of the steroid solution on a piece of glass wool, followed by evaporation of the solvent. The reader is cautioned, however, that the application of this system would depend on the availability of gas chromatographs which allow interruption of the gas stream and introduction of individual samples. While sample introduction by solid injection would appear preferable, this particular system suffers from the fact that pressure re-equilibration may be too slow for good chromatography. Nevertheless, this objection is readily overcome by the elegant solid state injection recently described by Menini and Norymberski [15], and modified by us as shown on page 27.

Bailey [16], quite recently, modified the original method for the determination of corticosteroids. The procedure is designed for the measurement of free cortisol and cortisone and for the determination of several polar cortisol metabolites (20 β-hydroxycortisol, 20β-hydroxycortisone, tetrahydrocortisol, tetrahydrocortisone, and 6β-hydroxycortisone) in urine without preliminary hydrolysis. The major steps of the procedure are the separation of the urinary free steroid into four distinct zones on paper in a Bush B5 system, oxidation with sodium bismuthate, and gas chromatography on XE-60.

The specificity of the procedure is derived from the thin-layer separation of the corticosteroids (Table 17-2), the comparison of the

TABLE 17-2. R_f Values of Corticosteroids in Thin-Layer Silica Gel [chloroform : ethanol : water (174:26:2)] *

Corticosteroid	R_f
Cortisone	0.61
Prednisone	0.57
Cortisol	0.42
20β-Hydroxycortisone	0.37
Tetrahydrocortisone	0.35
6β-Hydroxycortisone	0.34
Tetrahydrocortisol	0.24
6β-Hydroxycortisol	0.22
20β-Hydroxycortisol	0.21
Cortolone	0.18
Cortol	0.11

*Reproduced by permission from E. Bailey, Gas Chromatography of Steroids in Biological Fluids, edited by M. B. Lipsett (Plenum Press, New York, 1965), p. 59, Table 1.

TABLE 17-3.   Relative Retention of Steroids*

Steroid	XE-60†	SE-52‡
11-Ketoetiocholanolone	0.466	0.677
11-Hydroxyetiocholanolone	0.603	0.925
17-KS of cortisone	1	1
17-KS of prednisone	1.18	1.06
17-KS of cortisol	1.55	1.50
17-KS of prednisolone	2.16	1.72
17-KS of 6β-hydroxycortisone	2.73	1.42
17-KS of 6β-hydroxycortisol	5.01	2.49
17-KS of 6β-hydroxycortisol TMSi	1.28	1.93
	XE-60¶	
Cholestane	1	
Androsterone TMSi	1.01	
Etiocholanolone TMSi	1.25	
Dehydroepiandrosterone TMSi	1.47	
11-Ketoetiocholanolone TMSi	3.21	
11-Hydroxyandrosterone TMSi	3.59	
11-Hydroxyetiocholanolone TMSi	4.36	

*Reproduced by permission from E. Bailey, Gas Chromatography of Steroids in Biological Fluids, edited by M. B. Lipsett (Plenum Press, New York, 1965), p. 63, Table 2.
†Retentions relative to 17-KS of cortisone = 16.0 min; 1% XE-60, 5 ft by 4 mm, 240°C, 27 p.s.i.
‡Retentions relative to 17-KS cortisone = 12.8 min; 2% SE-52, 5 ft by 4 mm, 210°C, p.s.i.
¶Retentions relative to cholestane = 5.6 min; 1% XE-60, 5 ft by 4 mm, 195°C, 27 p.s.i.

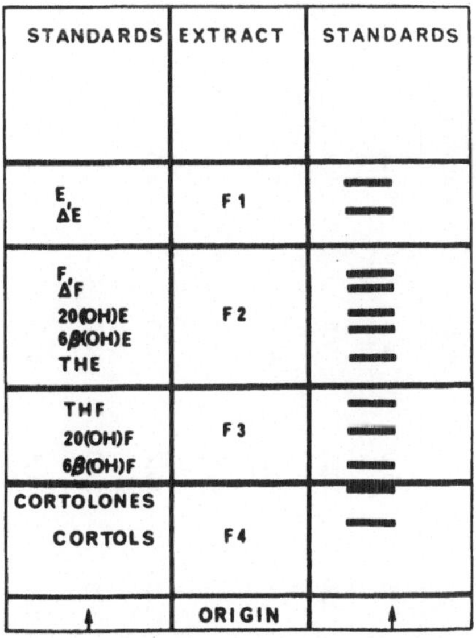

STANDARDS	EXTRACT	STANDARDS
E, ΔE	F 1	▬ ▬
F, ΔF 20(OH)E 6β(OH)E THE	F 2	▬ ▬ ▬ ▬ ▬
THF 20(OH)F 6β(OH)F	F 3	▬ ▬ ▬
CORTOLONES CORTOLS	F 4	▬ ▬
↑	ORIGIN	↑

Figure 17-3. Thin-layer chromatogram of standards and extract in the chloroform: ethanol: water (174:26:2) system showing the division of the urine extract into four fractions.

retention times of the peaks from extracts with authentic compounds (Table 17-3) after side-chain cleavage, and the increase of peak height without peak splitting or change in symmetry after addition of pure compounds to extracts. Further evidence is presented by trapping the peak materials and rerunning them on an SE-52 column, and by comparison of the retention times of the trimethylsilyl ethers, of those substances capable of forming them, with authentic materials.

Particular care must be taken in separating the four fractions on the thin-layer chromatogram (Figure 17-3) since the identical compound is formed from more than one fraction during side-chain cleavage and the assignment of the original structure is based entirely on the position of the precursor on the thin-layer chromatogram. Addition of known concentrations of prednisone and tritiated cortisol to five urine specimens resulted in the recovery, after completion of the total procedure, of 79.8% of prednisone (as measured by gas–liquid chromatography) and 79.9% of cortisol (as measured by radioactivity) just prior to GLC. In a pool of 4 hr urine specimens the determination of the seven corticosteroids in six equal aliquots showed a very high degree of reproducibility of the procedure (Table 17-4).

Figures 17-4, 17-5, and 17-6 show the results obtained following GLC of three fractions of a normal urine extract. Fractions 1 and 2 were chromatographed directly after bismuthate oxidation, as was Fraction 3 (Figure 17-6). It should be noted, however, that the

TABLE 17-4.   Replicate Determinations of Corticosteroids in Combined 4 hr Urine Specimens
(μg per 450 ml aliquot)*

Aliquot	1	2	3	4	5	6	Mean	Standard deviation †
Tetrahydrocortisone	5.6	4.6	—	6.0	5.8	5.3	5.5	0.6
Tetrahydrocortisol	9.8	9.3	11.2	8.0	8.6	10.1	9.5	1.24
Cortisone	37.3	35.8	34.0	34.7	32.1	33.8	34.6	2.3
Cortisol	25.0	20.1	19.8	22.5	22.7	21.2	21.9	1.7
20β-Hydroxycortisone	22.1	19.1	22.7	21.0	21.3	19.8	21.0	1.5
20β-Hydroxycortisol	51.3	49.8	57.2	52.2	56.3	54.3	53.3	3.6
6β-Hydroxycortisol	39.2	39.1	45.1	39.5	35.7	41.6	40.1	3.7

*Reproduced by permission from E. Bailey, Gas Chromatography of Steroids in Biological Fluids, edited by M. B. Lipsett (Plenum Press, New York, 1965), p. 64, Table 3.
†Corrected for small numbers.

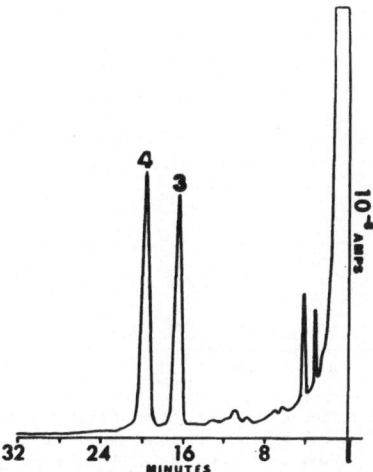

Figure 17-4. Gas Chromatogram of Fraction F1 after bismuthate oxidation ($^1/_{20}$ of a normal 4 hr urine). Peak 3: 17-KS of E derived from E. Peak 4: Operating conditions (A).

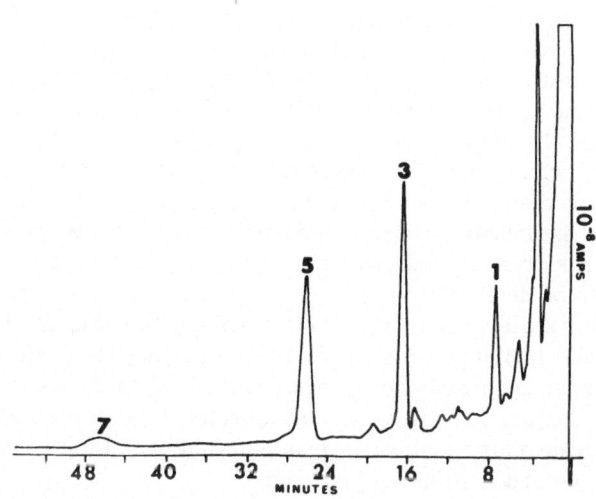

Figure 17-5. Gas chromatogram of Fraction F2 after bismuthate oxidation ($^1/_{10}$ of a normal 4 hr urine). Peak 1: 11-ketoetiocholanolone derived from THE. Peak 3: 17-KS of E derived from 20 β (OH)E. Peak 5: 17-KS of F derived from F. Peak 7: 17-KS of 6 β (OH)E derived from 6 β (OH)E. Operating conditions (A).

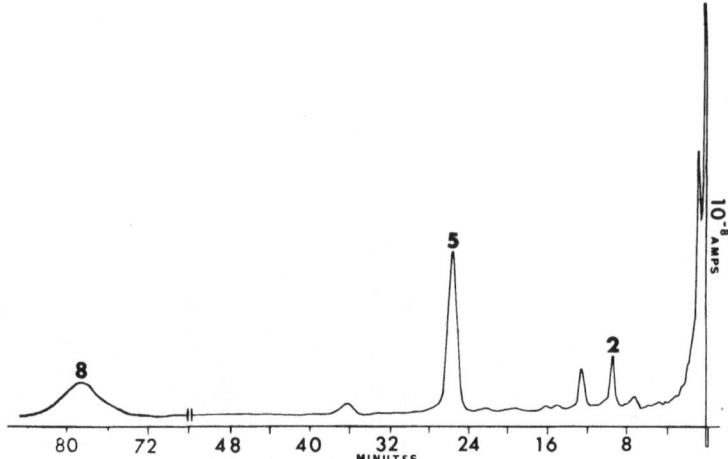

Figure 17-6. Gas chromatogram of Fraction F3 after bismuthate oxidation ($^1/_{20}$ of a normal 4 hr urine). Peak 2: 11-hydroxyetiocholanolone derived from THF. Peak 5: 17-KS of F derived from 20β(OH)F. Peak 8: 17-KS of 6β (OH)F derived from 6β (OH)F. Operating conditions (A).

derivative of 6β-hydroxycortisol had a retention time of nearly 80 min under those conditions. This is not only wasteful of time, but generally results in a significant decrease in peak height, possibly preventing its observation in the chromatogram. It is as a result of this finding that the authors have recommended the formation of the trimethylsilyl ether derivative of Fraction 3 (Figure 17-7) which results in a much more rapid elution of this compound. It is quite interesting to note that despite the obvious complexity of the procedure and the number of compounds which can be analyzed in a single specimen, it was possible for them to measure three urine specimens each day. As yet, little information about the sensitivity of the procedure is available, but it is stated that as little as 2 μg per urine specimen of cortisol can be determined.

Murphy, Bailey, and West [17] have applied this method to the study of the inhibitory action of 16-methyleneprednisolone and in doing so have extended the general procedure to include the corti-costeroid conjugates, although measurement of these substances greatly increases the number of manipulations and time consumption, as well as the expense, of the procedure.

In the three cases studied, the conjugated metabolites of cortisol reacted in exactly the same way as the free metabolites, suggesting that under some circumstances at least it may be entirely adequate to utilize the shorter assay procedure measuring only the free steroids.

We would be remiss in not calling the reader's attention to a recently published method which provides for a relatively simple method for screening of urines to determine 17-ketogenic steroids. With the use of borohydride reduction, side-chain cleavage, and further oxidation with *tert*-butylchromate, Menini and Norymberski [15] developed a procedure which gives an approximate picture of the hormonal status of the individual with respect to androgens and corticosteroids. Despite its obvious elegance, the method suffers from the drawback that a number of closely related substances are reduced to common denominators and it is not possible to tell the specific steroid which may be deficient or in excess. Indeed, where the balance of two substances differing only in their reduction–oxidation state is of critical interest (e.g., ratio of androstenedione to testosterone), only values for the combined ketonic base compound would be obtained.

EXPERIMENTAL

Method GC-I (Rosenfeld)

Preliminary Fractionation.   A portion of the neutral extract* from enzyme-hydrolyzed urine, corresponding to from 20 to 50% of a day's output, is applied in a thin line along the origin of a 20 by 20 cm plate coated with a 0.5 mm thick layer of Silica Gel G. At each side of the line, the plate is spotted with a mixture of cortol and cortolone. The

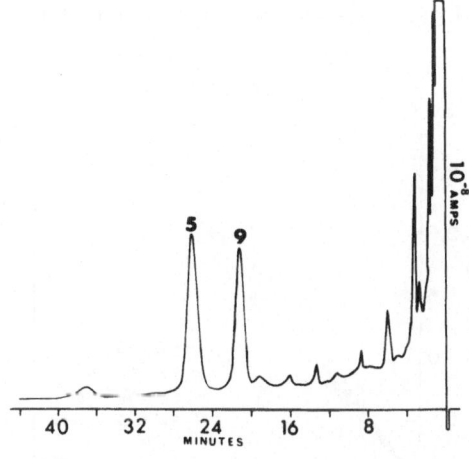

Figure 17-7. Gas chromatogram of Fraction F3 after bismuthate oxidation and trimethylsilyl ether formation ($^1/_{20}$ of a normal 4 hr urine). Peak 9: TMSi derivative of the 17-KS of 6β(OH)F derived from 6β(OH)F. Peak 5: 17-KS of F derived from 20(OH)F.

*Prepared exactly as outlined in method PT-I, page 203.

chromatogram is developed in ethyl acetate:cyclohexane (7:3) until
the solvent front moves 10 cm from the origin (20 min). The plate
is air dried and the area which contains the cortols and cortolones
is located by spraying the marker portion with phosphomolybdic acid
while carefully covering the center of the plate. The appropriate area
is removed from the plate, transferred to a small Soxhlet cup, and
continuously extracted overnight with refluxing methanol. Methanol
is removed *in vacuo* and the residue is converted to the TMSi ether
derivatives for analysis.

Derivative Preparation. The extract from thin-layer chromatography
is placed in a 3 ml centrifuge tube to which is added 0.8 ml of pyridine, $80\,\mu$l of hexamethyldisilazane, and $16\,\mu$l of trimethylchlorosilane. The mixture is sealed with a Teflon cap, agitated for a minute,
and allowed to remain overnight; the reagents are then removed under
a nitrogen stream at 60°. From 0.50 to 2.00 ml of chloroform containing androstane-3,11,17-trione (0.25 mg/ml, internal standard),
depending on the $C_{21}O_5$ content expected, is added, the contents are
thoroughly mixed and then centrifuged for 5 min. Portions are injected into the gas chromatograph, avoiding withdrawal of the packed
precipitate into the syringe.

Gas Chromatography. From 2 to $8\,\mu$l of the chloroform solution is
injected with a $10\,\mu$l Hamilton syringe onto a QF-1 column (3 wt.% on
100-140 mesh Gas Chrom P, 1.8 m by 4 mm; $T = 231°$; 30 psi argon
pressure). Weight vs. peak height curves (Figure 17-8) are constructed by injections of standards interspersed with the sample
runs. A typical gas chromatogram of a derivatized extract is shown

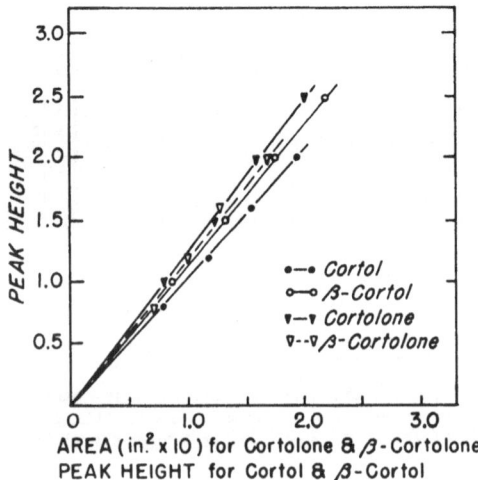

Figure 17-8. Standard curves for the trimethylsilyl ethers of cortols and cortolones.

in Figure 17-2. Quantitation is carried out by direct peak height comparison of each peak from the extract with the appropriate standard.

Method GC-II (Bailey)

Materials

Silica Gel G. (Merck, A.G. Darmstadt).

Ethyl Acetate (reagent grade) and Dichloromethane. Distilled twice.

Methanol and Ethanol (analytical grade). Refluxed with 2,4-dinitrophenylhydrazine and concentrated HCl, filtered and distilled twice.

Sodium Bismuthate (Reagent grade)

Hexamethyldisilazane

Trimethylchlorosilane

$1,2-^3H$-Cortisol. Repurified before use by thin-layer chromatography.

Scintillation Fluid. Prepared by adding 4 g of PPO and 100 mg of POPOP to 1 liter of toluene.

Extraction. A 4 hr morning specimen is extracted (unless ACTH or cortisol is given) after the internal standards (prednisone and $1,2-^3H$-cortisol) are added to the urine for the estimation of recovery and for quantitation. Sodium sulfate, 20 wt.%, is added to the urine and the mixture is extracted twice with equal volumes of ethyl acetate. The extracts are washed twice with $\frac{1}{20}$ volume of 1 N NaOH containing 15 wt.% of sodium sulfate, and once with $\frac{1}{20}$ volume of 0.5% acetic acid containing 15 wt.% of sodium sulfate. The combined extracts are dehydrated over sodium sulfate and taken to dryness in a rotary evaporator. The extract is next partitioned between petroleum ether (boiling point 100° – 120°C) and 70% aqueous methanol.

Thin-Layer Chromatography. The urine extract is dissolved in a little methanol and applied as a band 3 cm from the edge of a 20 × 20 cm glass plate coated with Silica Gel G (0.5 mm thickness). The following standards are applied on either side of the extract: cortisone [E], cortisol [F], tetrahydrocortisone [THE], tetrahydrocortisol [THF], and 20β-hydroxycortisol [20β(OH)F].

The chromatogram is developed in a fully saturated tank in the system chloroform:ethanol:water (174:26:2) at 28°C. The solvents are allowed to ascend 15 cm from the origin (approximately 45 min). The plates are dried in air at room temperature and the standards sprayed with an 0.2% solution of rhodamine 6G in ethanol. The Δ⁴-3-ketosteroids are located as dark bands on a yellow background under ultraviolet light (2530 A). Corticosteroids with a reducing side chain are located by a second spray of 0.1% alkaline solution of blue tetrazolium. The urine extract is then divided into 4 fractions which contain the areas located as follows:

Fraction (1), from just above E to just above F.
Fraction (2), from just above F to just above THF.
Fraction (3), from just above THF to just below 20β(OH)F.
Fraction (4), from just below 20β(OH)F down to and including the origin.

Each fraction is removed with an extractor or by scraping the adsorbent into a sintered glass funnel, and eluted with 8 ml of ethanol. The eluates then are taken to dryness.

Oxidation. The dried extracts from Fractions 1, 2, and 3 are shaken in the dark for 1 hr with 1 ml of 15% acetic acid and 25 mg of sodium bismuthate. Then, 2 ml of saturated $NaHCO_3$ are added and the reaction mixture is extracted twice with 20 ml of dichloromethane, washed with 4 ml of N NaOH and 4 ml of distilled water, dried over sodium sulfate, and evaporated under reduced pressure. Fraction 4 is oxidized with periodic acid instead of sodium bismuthate in order to oxidize to 17-ketosteroids (17-KS) the corticosteroids with a glycerol side chain and not those with a dihydroxy acetone side chain. Each oxidized fraction is taken up into 1 ml of ethanol and suitable aliquots used for gas chromatography.

Preparation of Trimethylsilyl Ethers. Dried aliquots representing half the oxidized portion of Fractions 2 and 3 are dissolved in 0.2 ml of chloroform and 0.2 ml of hexamethyldisilazane, and a few drops of trimethylchlorosilane are added. The mixture is allowed to stand at room temperature overnight, or at 60°C for 1 hr, in a tightly stoppered tube. The excess reagents are removed under nitrogen and the TMSi ethers taken up into 0.5 ml of hexane. Suitable aliquots are used for gas chromatography.

Isotope Recoveries. To estimate losses during the extraction, TLC, and oxidation procedures, two aliquots of the final extract of Fraction 2 are taken for counting in a scintillation spectrometer.

Gas-Liquid Chromatography. The injection port of the gas chromatograph is modified to accommodate a flash heater so that a method of solid injection can be employed [15]. Columns, 5 ft by 4 mm ID, are packed with 100-120 mesh, acid-washed and siliconized diatomaceous earth coated with 1% XE-60. The column is conditioned for 48 hr at 250°C. (Its useful life is increased by reducing the temperature to 175°C when not in use.) The column is operated with a flow of 90 ml/min (about 27 psi) of carrier gas at 240° and 195°C (for TMSi ethers) with the flash heater at about 260°C.

Quantitation is carried out by comparing the peak areas (or peak heights) of each substance to that of the internal standard (prednisone) after establishing mass response ratios for each. Final computation of the excretion is based on further correction as established by measurement of radioactivity of the tritiated standard remaining in the extract.

REFERENCES

1. D. K. Fukushima, J. Biol. Chem. 212:449 (1955).
2. W. J. Reddy, D. Jenkin, and G. W. Thorn, Metabolism 1:511 (1952).
3. J. K. Norymberski, Nature 170:1074 (1952).
4. W. J. A. VandenHeuvel and E. C. Horning, Biochem. Biophys. Res. Commun. 3:356 (1960).
5. H. H. Wotiz, I. Naukkarinen, and H. E. Carr, Biochim. Biophys. Acta 53:449 (1961).
6. I. Merits, J. Lipid Res. 3:126 (1962).
7. G. W. Kittinger, Steroids 3:21 (1964).
8. M. A. Kirschner and H. M. Fales, Anal. Chem. 34:1548 (1962).
9. H. Gottfried, Steroids 5:385 (1965).
10. R. S. Rosenfeld, Steroids 4:147 (1964).
11. D. K. Fukushima, H. L. Bradlow, L. Hellman, B. Zumoff, and T. F. Gallagher, J. Biol. Chem. 235:2246 (1960).
12. B. Hurlock and P. Talalay, J. Biol. Chem. 227:27 (1957).
13. J. A. Luetscher and R. G. Gould, J. Chromatog. 13:350 (1964).
14. E. Bailey, J. Endocrinol. 28:131 (1964).
15. E. Menini and J. K. Norymberski, Biochem. J. 95:1 (1965).
16. E. Bailey, Airlie House Conference, Warrenton, Va., February 1965.
17. D. Murphy, E. Bailey, and H. F. West, Lancet 809 (1963).

Chapter 18

Determination of Urinary Estrogens
(in collaboration with Sati C. Chattoraj)

BIOCHEMISTRY AND EVALUATION OF METHODS

The estrogens constitute a group of hormones, which with other hormones, take part in the development and maintenance of the female sex organs and the maintenance of the menstrual cycle and pregnancy. In the female the organs of origin are the ovary, placenta, and the adrenal cortex, and in the male the testes and adrenal cortex. Chemically, they are steroids having 18 or 19 carbon atoms, an aromatic A-ring and an hydroxyl group at C-3, imparting to them phenolic properties. Until 1955, only three classical estrogens—estrone, estradiol, and estriol—had been identified from human sources. Since then a dozen or more estrogens have been isolated and identified [1] (Figure 18-1). The clinical importance of the estimation of these hormones lies in the fact that these steroid hormones are known to be related to abnormal pregnancies, gynecological disorders, and some neoplasias. While an understanding of the relationship of the estrogens to such abnormal conditions requires the assessment of secretory and excretory rates of the hormones, methodological deficiencies (among others) for their evaluation has kept the exact nature of their function a mystery. The difficulty in assaying this group of hormones is largely due to the fact that except in pregnancy and in certain endocrinopathies, they are present in extremely small quantities in biological media while contaminated with large amounts of other steroidal and nonsteroidal substances. The discovery of nearly twenty metabolites in recent years has made the picture even more complicated.

Since 1930, many methods based on the chemical or biological properties of these hormones have been developed. In recent years the application of adsorption and column partition chromatography coupled with colorimetric and fluorimetric measurement have made possible the development of several very elegant procedures [2–4]. Nevertheless, none of them satisfy all the criteria necessary for a versatile and sensitive analysis. Apart from the adequacy of sensitivity, the clinical

225

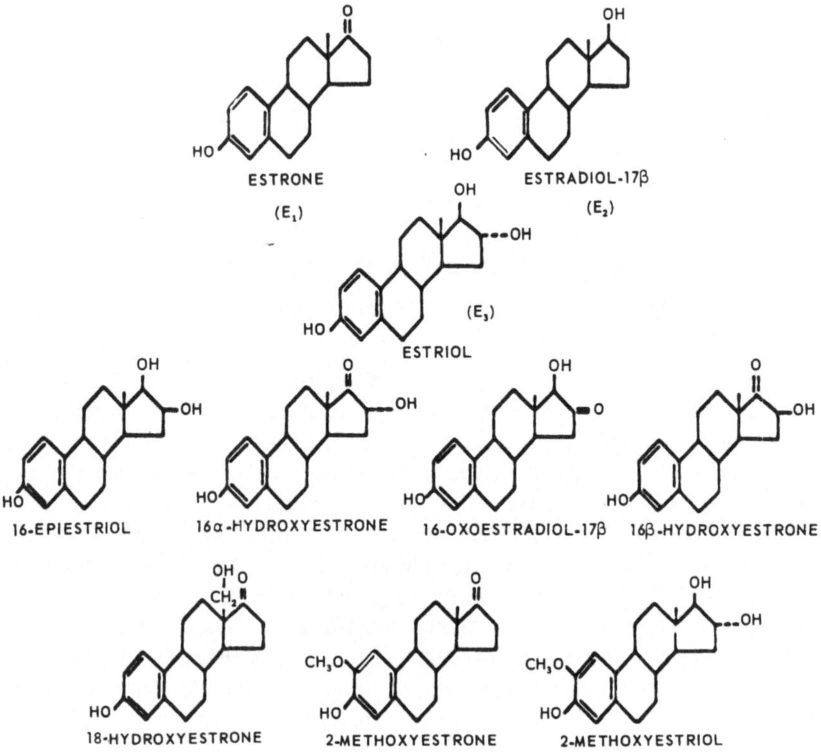

ESTRONE
(E₁)

ESTRADIOL-17β
(E₂)

ESTRIOL
(E₃)

16-EPIESTRIOL   16α-HYDROXYESTRONE   16-OXOESTRADIOL-17β   16β-HYDROXYESTRONE

18-HYDROXYESTRONE   2-METHOXYESTRONE   2-METHOXYESTRIOL

Figure 18-1. Chemical structures of some of the estrogens.

application of estrogen determination requires the analysis to be performed within a short period of time. The best available chemical methods are generally too time-consuming to play an important role in diagnosis or therapy. For a detailed discussion of the chemical determination of estrogens, the reader is referred to the excellent reviews by Preedy [5] and Bauld and Greenway [6].

Reports on the feasibility of gas chromatography [7, 8] and its application in the quantitative analysis of estrogens [9–11] have brought new techniques into play toward resolving this long standing problem. Following these earlier reports a number of procedures using gas chromatography have been published for the detection of estrogens from various biological media [12–15], but seldom have the quantitative aspects of this new tool been properly evaluated.

The suitability of different stationary phases, derivatives vs. free compounds, and the requirements of preliminary purification for actual reproducible analysis of estrogens in extracts often are not critically determined. Unfortunately, one forgets the fact that the experimental conditions found suitable for standard reference compounds and single isolated applications to biological material do not necessarily indicate their applicability in the routine quantitative analysis of urines or plasma. As a result the accuracy, precision, and specificity of gas chromatographic methos are frequently in doubt.

Before the detailed aspects of the procedure for the estimation of various estrogens are discussed, it is important to consider the different prerequisites necessary for the separation, detection, and quantitative evaluation of estrogens, particularly as applicable to their analysis by gas–liquid chromatography. For the sake of clarity these prerequisites may be divided into several categories: formation of derivatives, solvent purity, preparation of column, choice of stationary phase, and instrumental design. Several of these points have already been covered in an earlier chapter.

Choice of Derivative

Steroids with polyfunctional groups show strong irreversible sorption on the active sites of the column. Derivatives of a relatively nonpolar nature, having adequate volatility, not only overcome these effects but generally enhance separation. The first description of separation and analysis of steroid acetates was presented by Wotiz and Martin [8]. The use of trifluoroacetates and trimethylsilyl ethers in steroid separation was reported by VandenHeuvel et al. [16] and Luukkainen et al. [11], respectively.

A comparison of a number of important characteristics of such derivatives is shown in Table 18-1, along with data for the mixed derivatives obtained by prior methylation at position 3, and for the parent compounds. The major objections to the use of acetates raised by some investigators are the longer retention times requiring higher column temperatures, asymmetry of the peak, and poor resolution. Examination of Table 18-1 and some of the chromatographic reproductions in the main body of the paper [8] show that with a properly prepared column excellent symmetry can be obtained. With respect to the speed of elution, identical retention times for any single compound may well be achieved by increasing the temperature only slightly. Most of the stationary phases presently in use tolerate temperatures of 250°C; the silicone elastomer SE-30 has been used successfully at temperatures up to 300°C. The stability of a number of steroids during gas chromatography at these and higher temperatures has already been discussed. In fact, the only disadvantage that the acetate derivatives present is their ability to sorb on the column, requiring saturation prior to analysis. Their stability in solution, during gas chromatography, and during further purification after derivative formation has been adequately demonstrated. Furthermore, acetylation of all unhindered hydroxyl functions of the known estrogens can

TABLE 18-1. Properties of Different Derivatives*

Property	Acetates	TMSi	3-O-Me (as acetate or TMSi)	Free steroid
Symmetry	Excellent	Excellent	Excellent	Poor
Quantitativeness	95-100%	Presumed quantitative	Presumed quantitative	—
Reactivity	—OH of all estrogens	—OH of all estrogens	α-Ketols decompose epiestriols-multiple peaks	—
Stability in GLC	Proven for 10 estrogens	Presumed	Proven	Proven for E_1, E_2, E_3
Stability in solution	Excellent	Tendency to hydrolyze (moisture or acid)	Excellent (as acetate)	Excellent
Stability for further procedures	Excellent	Poor	Excellent (as acetate)	Excellent
Ease of preparation	Good	Fair	Poor	—
Comparative temperature of elution of E_1 for equal t_r (5.4 min) on a 3% SE-30 column	245°	239°	Unknown	232°
Nonadsorptivity on column	Fair	Very good	Unknown	Poor

*Reproduced by permission from H. H. Wotiz and S. C. Chattoraj, Gas Chromatography of Steroids in Biological Fluids, edited by M. B. Lipsett (Plenum Press, New York, 1965), p. 196, Table 1.

TABLE 18-2. Effect of Varying the Ratio of Acetic Anhydride/Pyridine on the Formation of Estriol Triacetate*

Ratio of acetic anhydride/ pyridine	Percentage yield range (Experiment No.)	Average	Remarks
1:5	56−65(5)	59.6	Presence of an earlier peak (t_r =12 min). 30% of the peak height of estriol triacetate (t_r =17.3 min)
1:2	68−75(5)	72.0	Early peak presents approximately 10% of the peak height of estriol triacetate
1:1	75−83(6)	80.0	Trace of early peak
2:1	85−94(5)	90.0.	No early peak discernible
5:1	95−101(7)	98.0	Single peak for estriol triacetate
8:1	94−100(5)	97.8	Single peak for estriol triacetate

*One or 0.5 μg of estriol was incubated with an 0.2 ml mixture of acetic anhydride and pyridine for 1 hr at 60°C. The reaction mixture was evaporated under N_2 and the residue was dissolved in 20 or 40 μl of acetone and 2 μl of the mixture was injected into the gas chromatograph. The percentage yield was calculated by comparing the peak height of the unknown with that of 0.05 μg of pure estriol triacetate. Column: 3% SE-30 on Diatoport S (80-100 mesh, alcohol washed), 6 ft by $\frac{1}{8}$ in. stainless steel. Column temperature: 250°C, 25 psi N_2, 5 psi H_2, 10 psi air.

TABLE 18-3. Acetylation Yields for Several Estrogens*

Compound	Amounts in μg reacted	Percentage yield range (Experiment No.)	Average
Estrone	0.5	98−102 (5)	99.6
Estradiol	0.5	96− 99 (5)	97.6
Estriol	0.5	96−102 (5)	98.2
16α -OH-Estrone	0.2	98−103 (5)	99.6
2-Methoxyestrone	0.2	96− 98 (5)	96.8
16-Ketoestradiol	0.5	98−104 (5)	99.8
16-Epiestriol	0.5	95− 98 (5)	96.4

*The acetylation and solvent partition was carried out as described under "Preparation of Derivatives."† The operating conditions for gas chromatography were the same as described in Table 18-2. †See page 262.

TABLE 18–4.   Comparison of Retention Times of Different Derivatives on Two Columns*

Compound	SE-30				QF-1		
	Free [a]	TMSi [b]	TMSi [c]	Acetate [d]	Free [e]	TMSi [f]	Acetate [g]
Estrone	4.9	5.5	6.1	7.4	11.0	7.7	7.0
Estradiol	5.1	7.2	7.8	10.8	6.8	3.7	7.0
2-MeO-Estrone	6.8	8.3	8.7	11.0	15.8	12.1	11.3
16α-OH-Estrone	7.2	10.4	10.5	15.4	15.4	8.5	16.7
16-Ketoestradiol	6.7	10.6	11.5	16.0	12.4	12.4	20.0
Estriol	10.5	14.6	14.2	20.0	18.1	6.6	15.3
16-Epiestriol	10.7	15.9	15.1	23.3	18.6	7.7	23.5

*Reproduced by permission from H. H. Wotiz and S. C. Chattoraj, Gas Chromatography of Steroids in Biological Fluids, edited by M. B. Lipsett (Plenum Press, New York, 1965), p. 198, Table 2.

†Taken from Luukkainen et al., Biochim. Biophys. Acta 62:153 (1962).

a 0.75% coating, column at 205°C.
b Same as above.
c 3% coating at 239°C.
d Same as above.
e 1% QF-1 at 195°C.
f Same as above.
g 4% QF-1 at 208°C.

be achieved quickly and in excellent yield (95–100%). This latter point should be particularly emphasized in view of the fact that Kliman reported that both estradiol and estriol were not acetylated in position 17 following reaction with acetic anhydride and pyridine. We have since carried out a reinvestigation of the acetylation of estriol and have found that the most advantageous ratio of acetic anhydride to pyridine for a quantitative yield is approximately 5:1.

Table 18-2 shows the results obtained following incubation of estriol with varying proportions of the two reagents. It can be seen quite clearly that 90% or better yields can be obtained with a 2:1 ratio, with approximately 98% yields at 5:1 and higher. With a ratio of 1:5, which is still considerably greater than that used by Kliman, as little as 60% acetylation of the 17-hydroxyl group occurred in 1 hr. Chromatography of this mixture yielded two major peaks, the faster of the two appearing with a retention time to be expected for estriol diacetate. Table 18-3 further verifies the quantitativeness of acetylation using a 5:1 ratio of reagents for a number of estrogens in submicrogram quantities.

Similar studies on the stability of trimethylsilyl ethers are still lacking in the literature. The TMSi ethers are presumed to be stable during gas chromatography and quantitativeness of reaction has been established only on the basis of gas chromatographic examination. Peak symmetry of these derivatives is excellent. Like the acetates, all the unhindered hydroxyls of known estrogens undergo reaction. Probably due to their greater volatility and bond shielding, sorption phenomena on the column are reduced for the TMSi ethers. On the other hand, stability in solution is more questionable, especially where any traces of moisture or acid are likely to be present. The tendency to hydrolyze spontaneously is considerable. It has been our experience that once these derivatives are formed, they frequently cannot be further purified without causing at least partial hydrolysis.

Use of the mixed derivatives, with the methyl group at position 3, while allowing the important advantage of purification through phase change prior to gas–liquid chromatography, has not been employed by us. This reaction would cause destruction of the a-ketols and was shown to produce more than one peak for both 16- and 17-epiestriols.

With respect to the suggestion that resolution of closely related estrogen acetates is inadequate, it must be realized that complete resolution of all the major estrogen metabolites, either as free steroids or derivatives from pregnancy urine, has so far not been achieved on one column.

Table 18-4 shows the actual retention times of the free steroids, acetates, and TMSi ethers on SE-30 and QF-1 columns. The solid lines connect compounds which cannot be resolved despite the small differences in retention time; the dotted lines link compounds which are only partially resolved with $R < 1.0$.* As can be seen, the best separations can be obtained for the acetyl derivatives on a QF-1 column, where only a single pair of compounds is not resolved. Similar difficulties in the separation of either free compounds, trimethylsilyl ethers, or acetates on polar or nonpolar columns can also be observed by examination of the present literature. In view of the number of other elegant techniques, such as thin-layer chromatography, for preliminary separation, this inability to achieve complete resolution of all estrogen metabolites is of relatively little consequence. It is entirely feasible to separate the estrogens into groups, each subgroup then being resolved by gas chromatography. This is not only necessary for separation of different metabolites, but also for removal of impurities prior to gas chromatography, and is highly desirable as a means of enhancing the specificity of the various methods.

Having evaluated the various qualities of these two types of derivatives, we have concluded that the use of estrogen acetates, particularly when applied in a routine analytical laboratory, is preferable.

The use of underivatized steroids [15] for direct analysis cannot be presently advocated since resolution is decreased, serious adsorption on many types of columns occurs, and severe peak skewing is generally obtained, resulting in uncertainty of specificity and poor accuracy and precision.

Determination of Estrogens in Pregnancy Urine

Rapid Determination of Estrone and Estriol in Crude Extracts of Late-Pregnancy Urine. The first practical application of gas chromatography to the analysis of hormones in biological media [17] described the determination of the classical estrogens in pregnancy urine. This method was found to be quite useful for the rapid determination of estriol in the second half of pregnancy. Its great advantage was the possibility of completing such an analysis in 2–3 hr.

Evidence for the accuracy of the procedure was obtained through analysis of the estrogens following their addition to urine samples. Mean recoveries of estrone, estradiol, and estriol of $75 \pm 6\%$, $80 \pm 4\%$, and $92 \pm 6\%$, respectively, were found. For quadruplicate determination of estriol in urines containing from 500 to 5000 μg, a reproducibility of $\pm 4.3\%$ was observed. When such a crude extract is utilized, serious doubts regarding the specificity of the procedure tend to occur. In order to ascertain that certain structures assigned to specific peaks contain primarily those designated, a large volume of urine was extracted and a phenolic extract prepared with minimum exposure to alkali. Following multiple injections into a gas chromatograph with a β-ionization detector and operation on argon gas, the respective peaks (Figure 18-2) were trapped in a test tube submerged in liquid

*See Chapter 3, page 10.

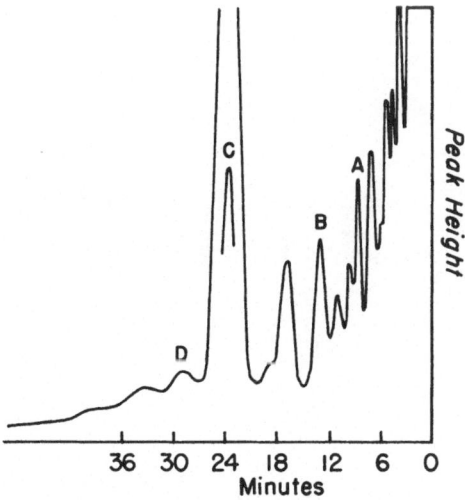

Figure 18-2. Chromatogram of a phenolic third trimester pregnancy urine extract (1/2050 aliquot) on a 3% SE-30 column.   A: estrone acetate; B: estradiol diacetate; C: estriol triacetate; D: 16-epiestriol triacetate. Column: 260°C; 14 psi. N_2.

nitrogen. Rechromatography of individual peaks with further trapping resulted in an overall recovery of 68% of the original material. Table 18-5 shows the data obtained to verify the nature of the compounds designated as estrone, estradiol, estriol, and 16-epiestriol. Adequate evidence for specificity has been obtained by means of paper chromatography, countercurrent distribution, and infrared data, and the peaks designated as estrone, estradiol, estriol, and 16-epiestriol are largely those compounds.

Figures 18-3 and 18-4 show the infrared spectra obtained from the two fractions designated as estrone acetate and estriol triacetate. Spectra for the fractions expected to contain estradiol diacetate and 16-epiestriol triacetate indicate the presence of both these substances, but the unduly large concentration of other contaminants makes the spectra unsuitable for publication, since this method is only recommended for the measurement of estrone and estriol. Since much of our early work was done on urines obtained from only a few individuals, it was not noted until sometime later that a number of specimens contain an apparently phenolic impurity with a retention time very nearly that of estradiol. Figure 18-5 shows a chromatogram on an SE-30 column with approximately 3300 theoretical plates yielding a clean estrone and estriol peak but showing the presence of the unknown impurity.   Estimation of estradiol under these circumstances is of course impossible.

It would appear therefore that use of crude extracts even in pregnancy urines should be relied on only for the rapid estimation of

TABLE 18-5. Characterization of Pregnancy Estrogens After Gas–Liquid Chromatography*

Peak No.	Suggested compound	Thin-layer chromatography (R_f)						Paper chromatography				Infrared spectroscopy	Analytical GLC t_r (min)	
		System I[a]		System II[b]		System III[c]		System A[d]		System B[e]			Authentic	Unknown
		Standard	Unknown	Standard	Unknown	Standard	Unknown	Standard	Unknown	Standard	Unknown			
1,2,3	Not investigated – probably nonsteroidal													
4	Not investigated											Phenolic – no other oxygen		
5	Estrone acetate	0.90	0.89	0.52	0.52	0.29	0.28	0.67	0.67	28 cm	28 cm	Positive identification	9.7	9.7
7	Estradiol	0.63	0.64	0.24	0.24	0.71	0.70	0.45	0.45	12 cm	12 cm		13.7	13.7
8,9	Mixture of ketols	Not investigated										Steroidal phenol, ketone present		
10	Estriol triacetate	0.13	0.13	0.05	0.05	0.31	0.31	0.01	0.01	3 cm	3 cm	Positive identification	24.6	24.6
11	16α-Epi-estriol tri	0.30	0.30	0.07	0.07	0.30	0.31	0.11	0.11	5 cm	5 cm		29.0	29.0

[a]System I—benzene:ethyl acetate (1:1).
[b]System II—petroleum ether:dichloromethane:ethanol (10:9:1).
[c]System III—petroleum ether:methanol (9:1).
[d]System A—isooctane:toluene:methanol:water (25:75:80:20).
[e]System B—toluene:propylene glycol.
*Chromatograms developed in Systems I and II, and A and B were stained with Folin-Ciocalteau's reagent while for System III, sulfuric acid–ethanol reagent was used. Analytical GLC were performed on columns (6 ft by ⅛ in.) of 3% SE-30 coated on Anachrom ABS (80–100 mesh). Column temperature 240 C, 15 psi N_2, 4 psi N_2, 10 psi air. The peak numbers refer to those of Figure 16.

estrone and estriol. No evidence has been obtained by us as yet that interfering substances derived from drug metabolism or dietary factors produce false results.

Determination of Estrone, Estradiol, and Estriol Throughout Pregnancy. Because of the difficulties mentioned above and our previous inability to develop liquid–liquid partition systems which would remove such impurities and allow direct gas chromatography of extracts, preliminary cleanup of the phenolic fraction on an alumina column has been found to yield satisfactory results. The details of this procedure are discussed under the low-level assay. Figure 18-6 shows the chromatogram of a phenolic pregnancy urine extract, following acetylation, alumina chromatography, and gas–liquid chromatography on a QF-1 column. It is worth emphasizing the excellent purification that can be obtained by this simple and fast running alumina column prior to GLC. Other than estrone and estriol, no significant peaks are discernible, suggesting that the impurities present are small enough compared to the estrogens and thus remain virtually undetectable at an attenuation of 200× for estrone and estriol, and an attenuation of 20× for estradiol. Quantitative measurement of such peaks may be achieved with considerable accuracy even for estradiol which is superimposed on a descending baseline.

Evidence for the specificity of this particular assay as applied to pregnancy urine is obtained through collection of the proper aliquots from the alumina column, followed by gas–liquid chromatography on both polar (NGS), nonpolar (SE-30), and ketone-selective (QF-1) columns. Identical retention times for standards and the peaks from the alumina column eluates were observed. Further evidence was derived from the fluorescence spectra obtained in phosphoric acid solution. Figure 18-7 shows the excitation and emission peaks for standard estradiol and a sample collected from pregnancy urine; Figure 18-8 shows similar curves obtained for an authentic mixture of estrone and estriol as compared to an aliquot of the respective eluates from the alumina column. Although this procedure is somewhat more involved than the crude assay for estrone and estriol, it has been our experience that approximately 25 assays a week can be carried out by one technician.

Determination of Seven Estrogen Metabolites in Pregnancy Urine. The discovery during the last decade of a number of estrogen metabolites beyond the classical three has considerably complicated the analytical aspects of estrogen determination not only with respect to the separation of these compounds from each other and from interfering materials, but also because of the known lability and relative alkali insolubility of some of these compounds.

The ability to separate neutral substances from the urinary phenols has always been considered to be a great advantage in the

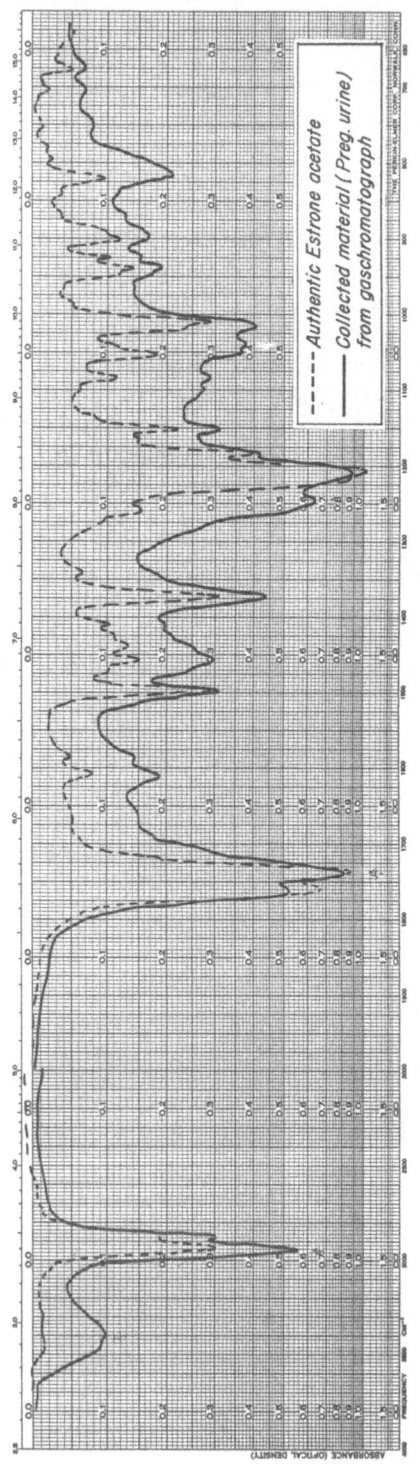

Figure 18-3. Infrared spectrum of estrone acetate isolated by gas chromatography from a pregnancy urine.

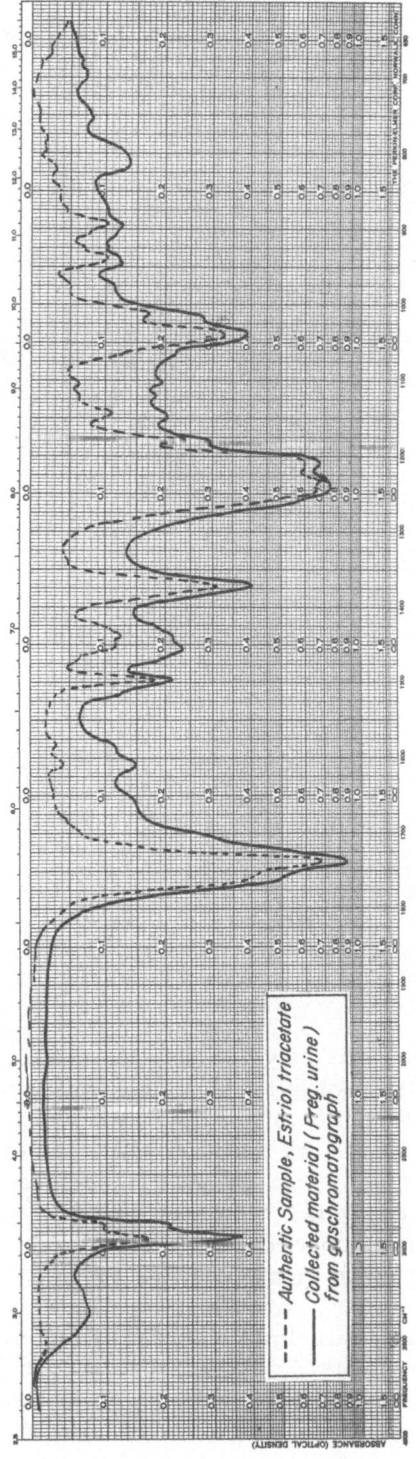

Figure 18–4. Infrared spectrum of estriol triacetate isolated by gas chromatography from a pregnancy urine.

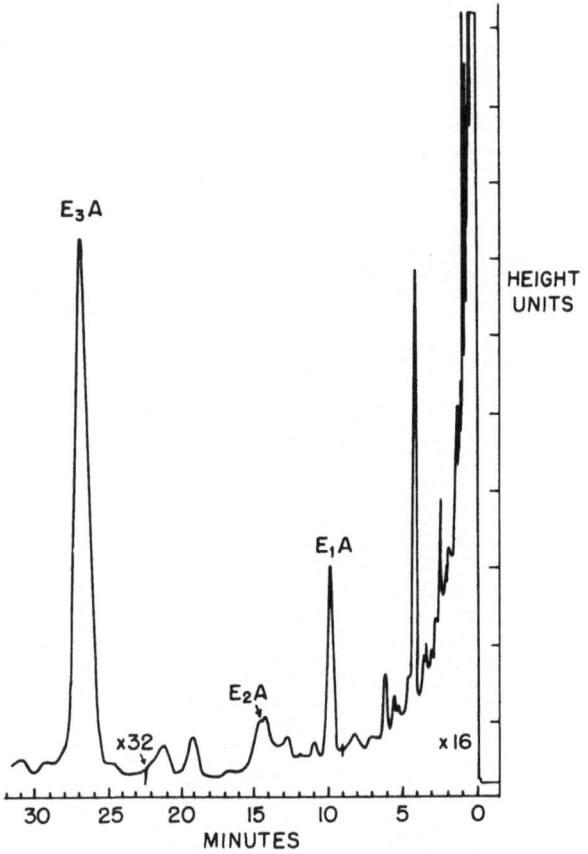

Figure 18-5. Gas chromatogram of an acetylated crude phenolic pregnancy urine extract on a 3.8% SE-30 column at 257°C and 55 ml/min He flow.

prepurification steps, markedly reducing the number of interfering chromogens. The low solubility of 2-methoxyestrone in alkali and the well-established lability of the ring-D-α-ketols not only make enzyme hydrolysis mandatory for such assay procedures, but also prevent the partitioning of the extracts between alkali and organic solvent. Consequently, it has been found necessary to employ other preliminary steps in the purification prior to gas chromatographic analysis of several of these compounds.

This procedure employs two different thin-layer chromatographic systems following enzymatic hydrolysis and extraction of the urinary estrogens. The purpose of the first system is twofold. First, separation of a large number of known urinary steroidal components from unknown colored substances can be achieved. Second, discrete zones

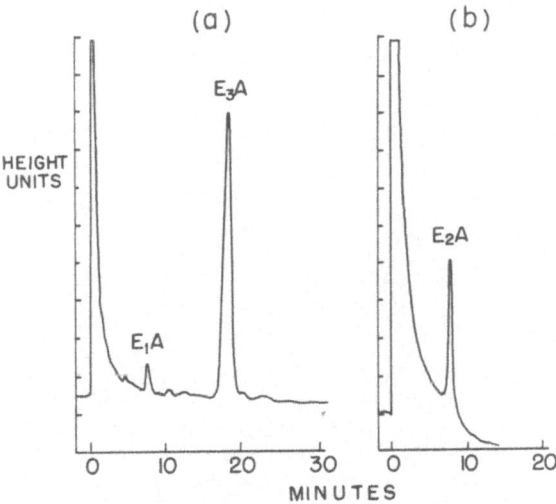

Figure 18-6. Chromatograms of a pregnancy urine extract following acetylation and alumina chromatography. (b): Analysis of estrone and estriol acetates on a 3% QF-1 column on 80-100 mesh Gas Chrom P; Attenuation: 200×; column temperature: 222°C; flow rate: 60 ml/min. (b): Estradiol diacetate fraction at attenuation 20×. Temperature and flow rate as above.

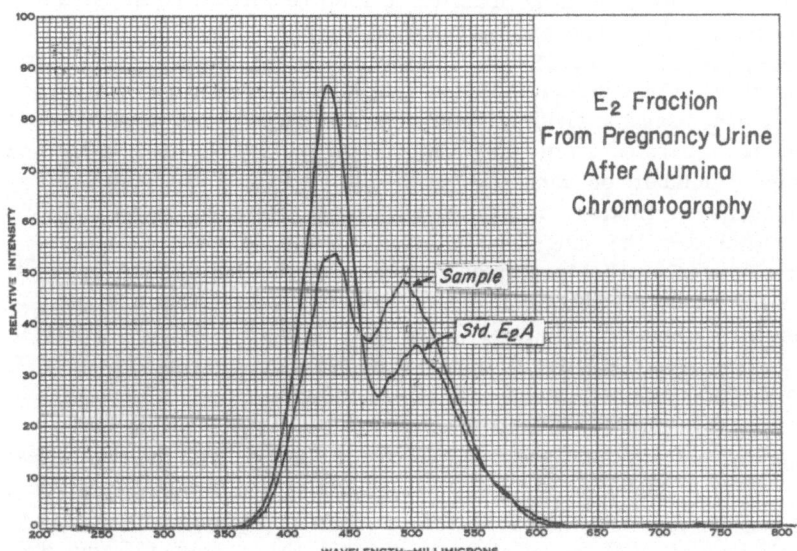

Figure 18-7. Comparison of fluorescence spectra of authentic estradiol diacetate and the sample obtained from pregnancy urine. Excitation peak: 425 mμ; emission peak: 500 m .

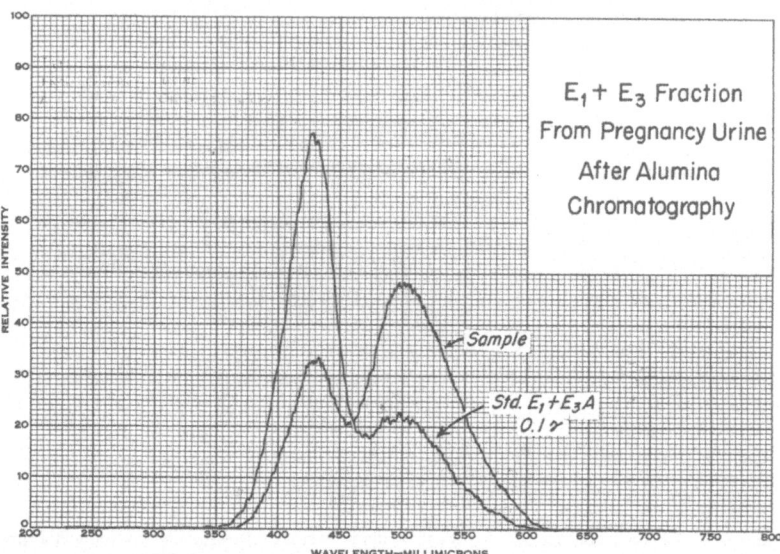

Figure 18-8. Comparison of fluorescence spectra of authentic estrone acetate and estriol triacetate with the corresponding sample obtained from pregnancy urine. Excitation peak: 425 mμ; emission peak: 500 mμ.

for 16-epiestriol, estriol, estrone, and 2-methoxyestrone, and for estradiol and the ring-D-a-ketols may be obtained. As pointed out earlier, the major difficulty in the separation of estrogen acetates on a QF-1 column lies in the inability to resolve estrone from estradiol. Separation of estrone and estradiol is readily achieved by using TLC System No. 1 (Table 18-6). Furthermore, the fractions containing estriol and 16-epiestriol appear sufficiently clean for direct gas chromatographic analysis. Similarly, estrone and 2-methoxyestrone may be separated from each other by gas chromatography as well as from a number of unknown peaks eluting faster than both of these estrogens. Elution and reapplication of Zone II, which contains estradiol and the ring-D-a-ketols, followed by development in a petroleum ether: dichloromethane: ethanol (10:9:1) system (Table 18-7), results in the separation of these three estrogenes from a number of interfering neutral steroids. Separation of 16 a-hydroxyestrone from 16-ketoestradiol finally is achieved by gas-liquid chromatography on a 0.5% EGA (stabilized) and 2.5% SE-30 column.

The recovery of various estrogens after the addition of 1 μg of each to 50 ml of urine varied from 72-91%, with the two lowest values (72% and 78%) found for 16-epiestriol and 2-methoxyestrone, respectively. The major losses of these two compounds appear to occur during thin-layer chromatography where only 81% and 82%, respectively, are recovered after chromatography alone, as compared to 90-97% for the other estrogens under discussion.

The linear response of the flame detector to six of these estrogens, varying from 0.02 μg to 2 μg and beyond, has been demonstrated. Table 18-8 shows the excretion pattern of these estrogens during a normal pregnancy. As a matter of expediency, the α-ketols have been determined on an SE-30 column as a combined peak.

TABLE 18-6. R_f and R_{E_1} Values of Free Steroids in System I [ethyl acetate:benzene (1:1)]*

Compound	Analyses	Mean R_f	Standard deviation	Mean R_{E_1}†	Standard deviation
Cortisol	7	0.092	±0.003	0.100	±0.015
Estriol	10	0.117	±0.010	0.119	±0.005
16-Epiestriol	11	0.289	±0.015	0.321	±0.032
11-Ketoetiocholanolone	6	0.294	±0.028	0.313	±0.020
Pregnanediol	6	0.347	±0.018	0.370	±0.011
Etiocholanolone and testosterone	6	0.477	±0.023	0.480	±0.014
16-Ketoestradiol-17β	12	0.633	±0.019	0.675	±0.017
Dehydroepiandrosterone, androstenedione, and androsterone	6	0.643	±0.023	0.610	±0.023
16α-Hydroxyestrone	6	0.658	±0.017	0.707	±0.004
Estradiol	11	0.679	±0.035	0.760	±0.017
Cholesterol	6	0.839	±0.031	0.904	±0.027
2-Methoxyestrone	7	0.842	±0.023	0.968	±0.004
16-Ketoestrone	6	0.881	±0.039	0.953	±0.019
Estrone	11	0.897	±0.056	1.000	

*Reproduced by permission from H. H. Wotiz and S. C. Chattoraj, Anal. Chem. 36:1466–1472 (1964), Table I. The system was left incompletely saturated.
†R_{E_1}= distance traveled relative to estrone.

TABLE 18-7. R_f and R_{E_1} Values of Free Steroids in System II [petroleum ether:dichloromethane:ethanol (10:9:1)]*

Compound	Analyses	Mean R_f	Standard deviation	Mean R_{E_1}[†]	Standard deviation
Estriol	8	0.050	±0.007	0.012	±0.001
Cortisol	6	0.052	±0.010	0.013	±0.002
16-Epiestriol	7	0.079	±0.007	0.202	±0.005
Pregnanediol	7	0.189	±0.040	0.468	±0.025
Estradiol-17β	9	0.215	±0.003	0.539	±0.028
16α-OH-Estrone	8	0.216	±0.003	0.538	±0.029
16-Ketoestradiol-17β	6	0.235	±0.013	0.581	±0.018
11-Ketoetiocholanolone	8	0.236	±0.021	0.592	±0.013
16-Ketoestrone	5	0.297	±0.028	0.737	±0.019
Etiocholanolone and testosterone	7	0.313	±0.027	0.798	±0.026
Dehydroepiandrosterone and androstenedione	8	0.355	±0.034	0.901	±0.036
Estrone	11	0.396	±0.028	1.000	
Cholesterol	6	0.544	±0.030	1.36	±0.01
2-Methoxyestrone	9	0.611	±0.025	1.54	±0.07

*Reproduced by premission from H. H. Wotiz and S. C. Chattoraj, Anal. Chem. 36:1466–1472 (1964), Table II. The system was fully saturated by lining the wall with filter paper.
† R_{E_1} = distance traveled relative to estrone.

TABLE 18-8. Estrogen Excretion During a Normal Pregnancy (μg)*

Date	Specimen No.	2-Methoxy-estrone	Estrone	Estradiol	Ring D-α-ketols‡	Estriol	16-Epi-estriol
1-27-61	1	51	575	N.D. †	558	9,604	133
2-16-61	2	130	520	N.D.	1109	16,750	145
3-6-61	3	243	789	N.D.	1412	23,219	533
4-6-61	4	185	1105	N.D.	1462	25,792	740
4-25-61	5	239	1084	N.D.	1527	27,988	§
5-3-61	6	144	1619	N.D.	1877	30,864	§
5-9-61			Parturition				

*Reproduced by permission from H.H. Wotiz and S.C. Chattoraj, Anal. Chem. 36:1466–1472 (1964). All values are corrected for recovery losses.
†Not detectable.
‡Adequate resolution could not be achieved either on thin-layer or gas chromatography. The value of ring D-α-ketolic estrogen is given with respect to 16α-OH estrone.
§Values were not determined because of an excessive spreading of the large amounts of estriol on the thin-layer plate obscuring the 16-epiestriol peak during GLC. The samples were lost by accident during a second TLC.

TABLE 18-9. Recovery Experiments in Percentages*

Experiment No.	Alumina column only:			Remarks
	Estrone	Estradiol	Estriol	
1	98.0	101.0	93.8	0.2 μg each of
2	90.0	96.4	90.7	estrogens were
3	93.0	88.5	95.0	added
4	90.0	94.3	98.0	
5	95.0	89.0	85.0	
	93.2 ± 3.5	93.8 ± 5	92.5 + 5	
	Overall recovery:			
1	68.5	72.0	80.6	0.5 μg E_1 and E_2
2	75.0	70.5	84.0	and 1 μg E_3 were
3	72.0	81.4	86.0	added to 1/20
4	70.0	75.0	84.0	of the total volume
5	76.0	79.0	79.0	of urine (male)
	72.4 ± 3.2	75.6 ± 4.4	82.8 ± 3.2	

*Reproduced by permission from H.H. Wotiz and S.C. Chattoraj, Gas Chromatography of Steroids, edited by M.B. Lipsett (Plenum Press, New York, 1965), p. 209, Table 6.

TABLE 18-10.   Precision Data on Duplicate Analyses*

Experiment No.	Estrone	Estradiol	Estriol
1. A	7.00	1.10	17.50
B	6.00	0.90	17.00
2. A	10.00	1.20	20.50
B	10.90	1.20	21.80
3. A	9.30	2.25	26.80
B	9.30	2.35	23.00
4. A	7.00	3.52	33.00
B	8.25	4.80	32.40
5. A	15.66	4.50	34.00
B	15.00	4.30	37.70
6. A	5.30	1.50	46.40
B	5.30	—	52.00
7. A	3.00	0.90	22.00
B	2.70	1.30	21.60
8. A	5.20	2.30	26.30
B	5.00	1.80	25.80
9. A	8.00	2.40	27.00
B	8.90	2.00	28.20
Variance of duplicates†	0.26	0.14	3.53

*Reproduced by permission from H.H. Wotiz and S.C. Chattoraj, Gas Chromatography of Steroids in Biological Fluids, edited by M.B. Lipsett (Plenum Press, New York, 1965), p. 210. Table 7.

†Calculated according to the formula $\dfrac{\Sigma(x-\bar{x})^2}{n-1}$.

We draw the reader's attention to the fact that we have been unable to determine any estradiol in this particular urine. Although a significant peak nearly corresponding in retention time to that of authentic estradiol was found in every sample when an SE-30 column was employed, chromatography of a second aliquot on the SE-30/EGA column showed no peak in the region of estradiol. It must be pointed out that this urine had been stored for well over two years in a freezer and periodically was defrosted and samples were removed from it. It is suggested that the storage and/or repeated defrosting and re-freezing of this urine are responsible for a molecular alteration of estradiol. We recently had opportunity to see this same phenomenon in another urine.

Evidence for the specificity of the reaction was primarily obtained through multiple thin-layer chromatography and color reactions, borohydride reduction of the ketols followed by TLC and finally GLC and mixed GLC [18].

Determination of Estrone, Estradiol, and Estriol During the Normal Menstrual Cycle. The great sensitivity of ionization detectors and the unique high resolution of the gas chromatographic column had raised the hope that the above procedure could be applied directly to urinary fractions containing very low levels of estrogens following simple preliminary separations. We have not, as yet, determined the lower limit of sensitivity of the method, but a preliminary survey on a few urines suggested that the assay would occasionally run into difficulty if more than 10% of the urine was used for analysis. The major difficulty here appeared to be a serious overloading of the thin-layer chromatogram. Consequently, we have reapproached this problem by attempting to develop a more sensitive procedure which may be utilized in a more routine manner even at the sacrifice, at the moment, of analyzing only the three classical estrogens.

The recovery of the three estrogens was studied, using the alumina column method outlined earlier, following column chromatography as well as after addition of steroid to a urine sample and carrying through the whole procedure (Table 18-9).

Although sufficient data for determination of the precision of the method are not yet available, a series of nine duplicate analyses of a nonpregnancy urine is shown in Table 18-10. A variance of 0.26, 0.14, and 3.53 for estrone, estradiol, and estriol, respectively, was found.

Tables 18-11 and 18-12 show the mean values obtained for the three classical estrogens, for several normal subjects after obtaining the specimens during the luteal and proliferative phases of the cycle. Table 18-11 shows the results obtained when Method E-III, which uses multiple thin-layer chromatography, was applied, but only estrone, estradiol, and estriol were analyzed. The results thus obtained compare favorably with the mean values found when another set of urines from normal volunteers was analyzed using Method E-IV, which utilizes the alumina column preliminary purification. The values obtained by the latter method are quite comparable to the data obtained from the more involved procedure and the small differences fall within the limit of experimental error.

Table 18-13 shows the mean values obtained during one complete menstrual cycle in a normal female. A plot of these data, as well as superposition of the pregnanediol excretion curve of this particular woman, are shown in Figure 18-9. The values obtained by the method described here appear to be in keeping with those reported in the past

TABLE 18-11.   Excretion of Estrogens in Adult Women (age 22 to 35 years) Using Method E-III

Subject No.	Day of cycle	Estrone μg per 24 hr	Estradiol μg per 24 hr	Estriol μg per 24 hr	Total μg per 24 hr
Proliferative Phase:					
1.	7th	3.50	1.60	15.00	20.10
	8th	4.35	3.70	15.00	23.05
2.	7th	6.30	2.00	20.95	29.25
	10th	9.00	6.25	20.45	35.70
3.	6th	7.00	2.45	19.35	28.80
	11th	12.00	4.35	18.00	34.35
4.	7th	4.00	3.00	12.00	19.00
	Average	6.59	3.30	17.25	27.18
Luteal phase:					
1.	16th	2.00	1.50	10.65	14.15
	17th	1.50	3.20	15.05	19.75
2.	15th	3.70	2.00	12.65	18.35
	21st	1.30	—	17.65	18.95
4.	15th	4.72	1.40	18.75	24.87
	19th	7.45	4.60	11.62	23.67
	20th	4.50	3.20	20.45	28.15
	Average	3.59	2.27	15.26	21.12

by application of the elegant procedure of Brown [2]. Pregnanediol was determined gas chromatographically by the method described in Chapter 14.

Typical chromatograms following GLC on SE-30 and QF-1 columns are shown in Figures 18-10 through 18-13. It can readily be seen that much better separation from interfering substances is obtained during chromatography of the estradiol fraction (Figures 18-10 and 18-12) on the fluoroalkyl silicone polymer (QF-1). Although no specific experiments have been carried out to determine the sensitivity of the procedure, it seems unlikely that concentrations of less than 0.5 μg per 24 hr can be measured with the present instrumentation.

Evidence for the specificity is inherent in the procedure and depends

TABLE 18-12.   Excretion of Estrogens in Different Phases of the Menstrual Cycle Determined by Method E-IV

Sample No.	Day of cycle	Estrone μg per 24 hr	Estradiol μg per 24 hr	Estriol μg per 24 hr	Total estrogens μg per 24 hr
Proliferative phase:					
1	6th	4.60	2.90	11.50	19.00
2	7th	3.60	1.20	16.50	21.30
3	8th	8.70	6.20	10.00	24.90
4	9th	5.60	2.50	12.22	20.32
5	10th	6.30	1.00	22.65	29.95
6	11th	4.00	—	25.00	29.00
7	12th	4.00	3.20	32.50	39.70
	Average	5.25	3.00	18.62	24.90
Luteal phase:					
8	15th	2.30	3.50	16.00	21.80
9	16th	4.00	2.60	7.30	13.90
10	17th	3.26	2.40	4.60	10.26
11	18th	3.26	2.40	21.40	27.06
12	19th	1.60	—	10.35	11.95
13	20th	5.20	1.35	28.05	34.60
	Average	3.54	2.45	14.61	19.93

to a large extent on the elution pattern from the alumina column and on multiple gas chromatographic determinations on polar (NGS), nonpolar (SE-30), ketone-retentive (QF-1), and mixed phase (NGS/SE-30) columns.   Identical retention times for the peaks from the extract and pure estrone, estradiol, and estriol were observed.

Further, saponification of the fractions of the acetylated extract and subsequent formation of the TMSi ethers resulted in identical t_r values for the respective estrogens and authentic estrogen derivatives on a QF-1 column.   Fluorescence spectroscopy of the eluate corresponding to estradiol produced the peak shown in Figure 18-14; the portion of the chromatogram corresponding to the mixture of estrone and estriol yielded the spectrum shown in Figure 18-15.

TABLE 18-13. Excretion of Estrogen in the Normal
Menstrual Cycle

Day of cycle	Estrone μg per 24 hr	Estradiol μg per 24 hr	Estriol μg per 24 hr	Total
6	6.50	1.00	17.30	24.80
7	10.40	1.20	21.10	32.70
8	9.30	2.30	24.90	36.50
9	7.50	4.20	32.7	44.40
10	10.11	2.50	27.00	39.61
11	15.30	4.40	35.80	55.50
12	5.30	1.50	49.20	56.00
13	6.40	1.60	21.20	29.20
14	2.85	1.10	21.80	25.75
15	5.10	2.00	26.00	33.10
16	8.40	2.20	27.50	38.10
17	4.40	0.95	22.00	27.35
18	7.80	3.25	18.00	29.05
19	4.00	2.50	12.40	18.90
20	3.60	—	13.00	16.60

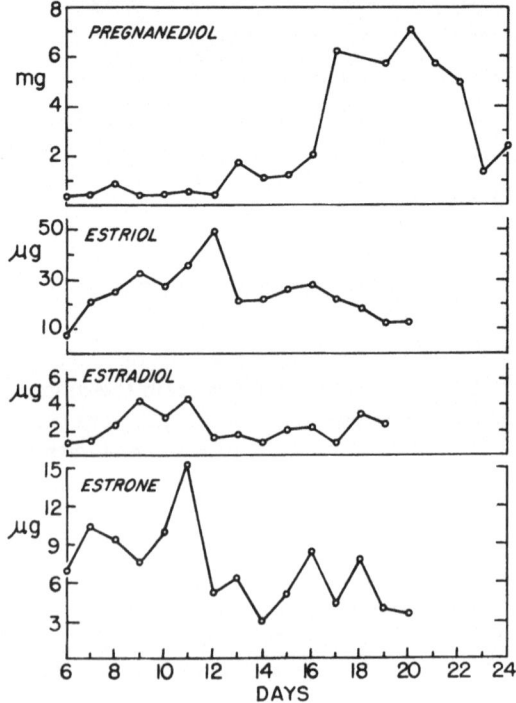

Figure 18-9. Urinary excretion of pregnanediol and estrogens during a normal menstrual
cycle as determined by GLC methods P-II and E-IV.

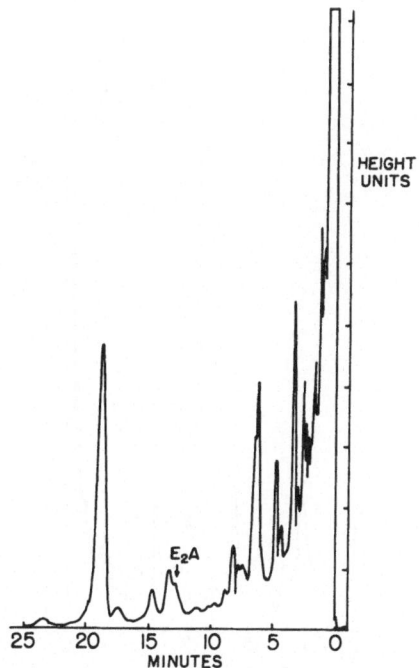

Figure 18-10. Gas chromatogram of the E₂A eluate (nonpregnancy urine) from the alumina column on a 6 ft × 4 mm, 3% SE-30 column.

Further evidence for the specificity of the method was based on the quantitative determination of individual fractions on a variety of columns. It may be considered to be fairly unlikely that two substances, even though very closely related, would display partition coefficients completely identical when distributed between a gas and a variety of stationary phases (polar, nonpolar, and ketone-selective) as the free compounds as well as two types of derivatives. From Table 18-14 it can be seen that the quantitative measurement of the peaks with retention times identical to the authentic estrogens gave fairly consistent results when chromatographed on different stationary phases. Following saponification of the acetylated fractions and subsequent formation of the TMSi ethers, identical retention times for the peaks from nonpregnancy extracts and authentic estrogen derivatives on a QF-1 column were obtained. If an approximate loss of 25% during saponification and TMSi ether formation is assumed, the results are quite compatible with the data obtained using acetyl derivatives.

Table 18-15 is a compilation of the reliability criteria of several of the presently accepted methods as well as two of the gas chromatographic procedures described by us. It would appear that the

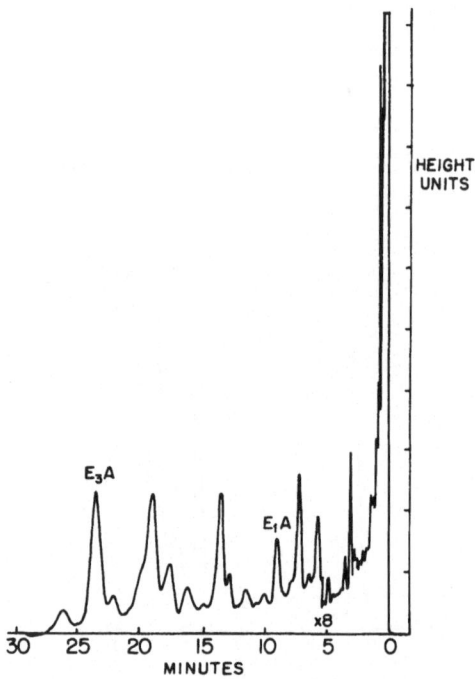

Figure 18-11. Gas chromatogram of the E_1A and E_3A eluate (nonpregnancy urine) from the alumina column on a 6 ft × 4 mm, 3% SE-30 column.

accuracy, precision, and specificity of the gas–liquid chromatographic procedures compare favorably with that of the other methods. The sensitivity and speed of estimation of the GLC methods are quite superior to that of the classical procedure. It is important, however, to point out that despite the greater rapidity, the precision of the GLC method appears to be better, although the accuracy is somewhat lower than reported by Brown, Bauld, and Preedy and Aitken. Moreover, the sensitivity for the GLC methods appears to be 10 to 20 times as great as for the classical procedures although much more data must be accumulated for complete validification.

Two avenues allowing possible improvement in the sensitivity of the method are presently under investigation. The utilization of solid injection, which would permit the introduction of all the final extract into the gas chromatograph, might allow a nearly tenfold increase in sensitivity.

Nevertheless, the anticipated advantage here would be relatively limited and other avenues of approach would seem more rational. Of particular interest in this respect is the use of derivatives which will allow their much more sensitive detection in an electron capture detector.

Clark and Wotiz [19] described a new derivative formed by the reaction of hydroxyl groups with heptafluorobutyric anhydride. These compounds show not only excellent electron capturing properties, but also allow good resolution of various estrogens. They have the further advantage of being extremely volatile, allowing elution temperatures below 200°C, a condition necessary for the safe maintenance of the tritium foil in the electron capture detector.

Figure 18-16 shows the separation and measurement of estrone and estradiol as the heptafluorobutyrates in concentrations of 0.1 ng each. As expected, estrone produces a considerably smaller peak since it reacts with only one heptafluorobutyryl group. Estradiol appears much faster in the chromatogram because of the presence of two such groups, thereby enhancing the volatility of the compound.

Methods for the quantitative formation of the heptafluorobutyrates of estrone, estradiol, and estriol have recently been worked out in the

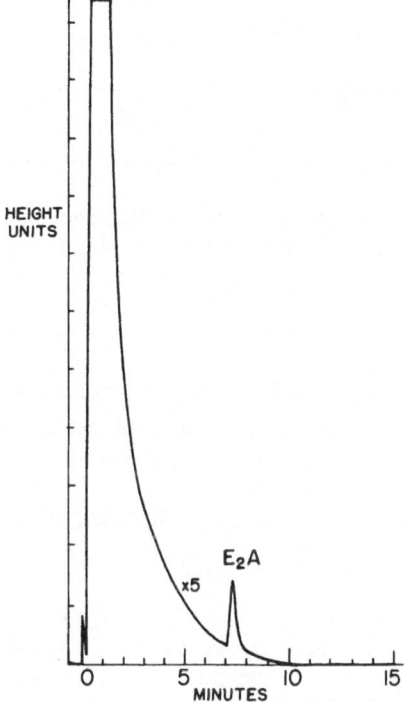

Figure 18-12. Gas chromatogram of the extract shown in Figure 18-10 on a 3% QF-1 column.

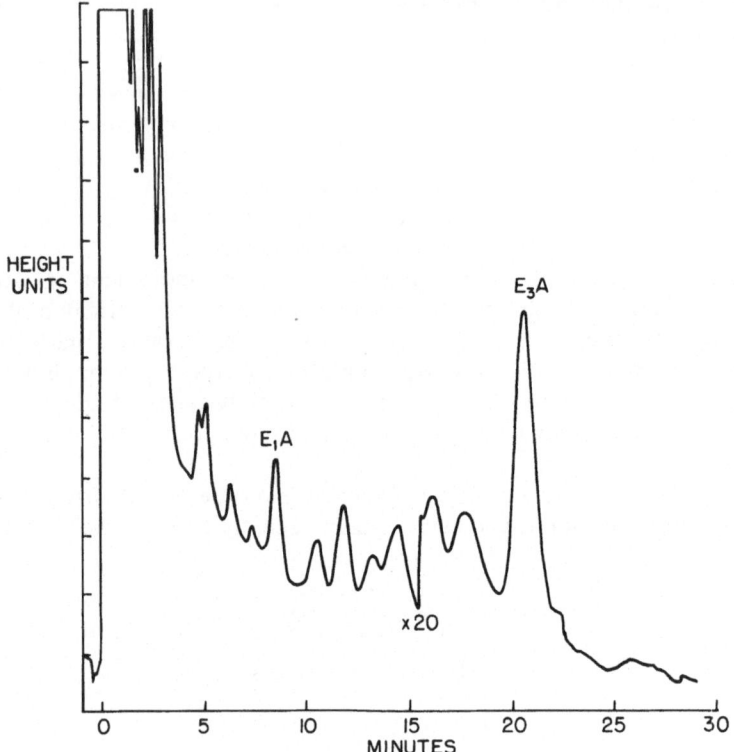

Figure 18-13. Gas chromatogram of the extract shown in Figure 18-11 on a 3% QF-1 column.

laboratory.   It is hoped that the application of these procedures to highly purified extracts of urinary and plasma estrogens will allow their sensitive determination in the future.

EXPERIMENTAL
Materials and Procedures

All reagents used were of analytical grade unless otherwise stated. The distillation of solvents was carried out in all glass apparatus. (Method E-I does not require the extensive purification of solvents described below.)

Ether.   Shaken with 13% (w/v) $AgNO_3$ (60 ml/liter), followed by washing with $1 N$ NaOH (100 ml/liter) and water. The ether is freshly distilled before use.

Petroleum Ether (30°-60°), n-Hexane, and Benzene. Washed with concentrated H_2SO_4 (3 × 100 ml/liter) followed by $KMnO_4$ in 4 N H_2SO_4 (3 × 100 ml/liter, water (3 × 180 ml/liter), 8% $NaHCO_3$ (30 ml/liter), and finally water (150 ml/liter) to neutral. The solvents are dried over Na_2SO_4 and then fractionally distilled.

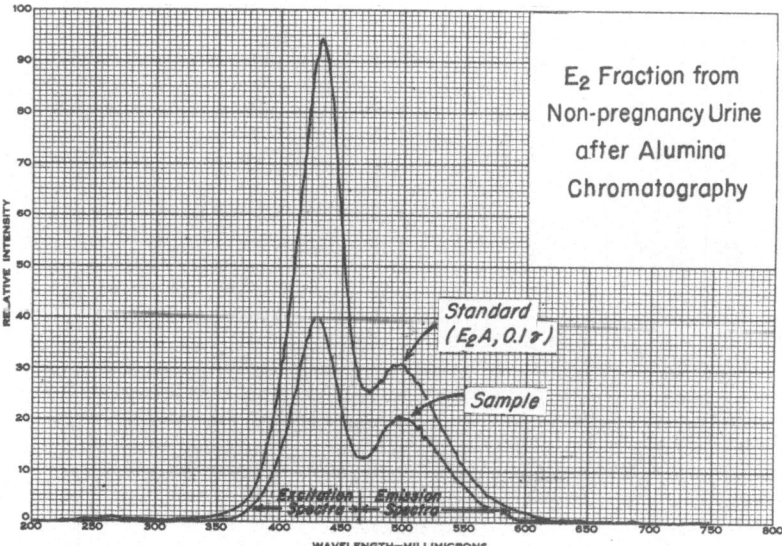

Figure 18-14. Fluorescence spectrum of E_2A fraction of nonpregnancy urine after alumina chromatography.

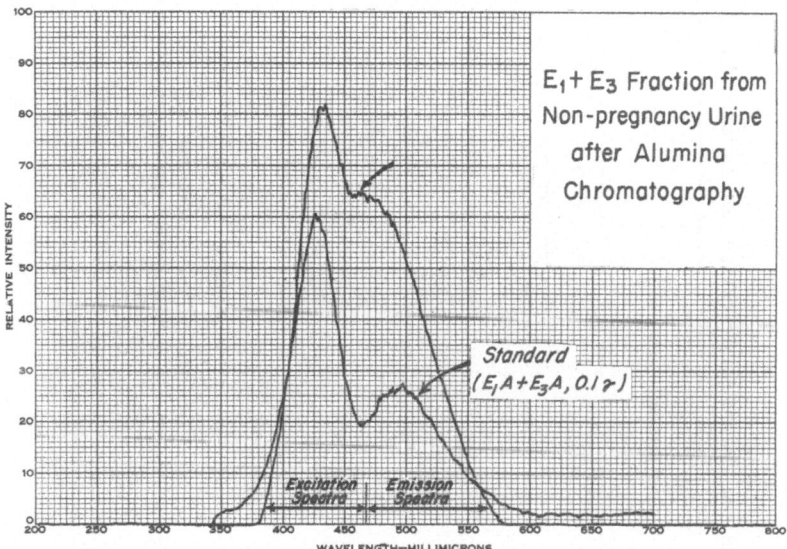

Figure 18-15. Fluorescence spectrum of E_1A and E_3A fraction of nonpregnancy urine after alumina chromatography.

TABLE 18-14. Quantitative Characterization of Estrogens by Gas Chromatography on Different Stationary Phases

Sample	Sample No.	Steroid	μg per 24 hr			
			SE-30	QF-1	XE-60	NGS and SE-30
Nonpregnancy Urine	CDS-3	Estrone acetate	8.00	9.30	8.90	8.95
		Estradiol diacetate	2.00	2.25	1.47	1.97
		Estriol triacetate	24.25	25.00	22.76	22.15
	CDS-10	Estrone ⎫		5.20		
		Estradiol ⎬ As acetates		2.00		
		Estriol ⎭		25.80		
		Estrone ⎫ As trimethylsilyl		4.00	After saponification of the	
		Estradiol ⎬ ethers		1.62	acetylated fractions, TMSi	
		Estriol ⎭		20.30	derivatives were formed	
						EGA and SE-30
Pregnancy Urine	AH-3	Estrone	789		720	750
		2-Methoxyestrone	243		230	220
		16α-Hydroxyestrone and 16-ketoestradiol	1,412		1,380	1,285 ⎱ 820
		Estriol	23,219		23,100	22,550
		16-Epiestriol	533		496	464 ⎰ 465

Dichloromethane. Shaken with anhydrous potassium carbonate (10 g/liter), filtered, and passed through a column of active silica gel followed by fractional distillation.

Ethyl Acetate. Refluxed with acetic anhydride (100 ml/liter) and concentrated H_2SO_4 (5 ml/liter) for 4 hr and distilled fractionally.

Ethanol and Methanol. Refluxed over solid NaOH and twice distilled under anhydrous conditions.

Pyridine. Refluxed over pellets of NaOH for 4 hr and fractionally distilled after addition of fresh pellets under anhydrous conditions.

Acetic Anhydride. Fractionally distilled under anhydrous conditions.

Aluminum Oxide. Neutral, activity grade 1 (M. Woelm, Eschwege, Germany). Deactivated by the addition of 5 ml of water per 100 g, shaken throughly to break up any lumps, and left overnight for equilibration in a tightly stoppered bottle. The activity is checked as follows:

The chromatographic column is partly filled with petroleum ether (according to the specifications given by Brown [2]) and 3 g of deactivated alumina is added in a thin stream so that it is freed from air as it settles. The surface of the alumina is leveled by tapping and is covered with a few glass beads. The rate of flow is adjusted to approximately 50 drops per minute. Five micrograms each of three estrogen acetates (estrone, estradiol, and estriol) in 25 ml petroleum ether is applied. Elution is carried out with the following solvent mixtures and 5 ml fractions are collected.

1. 25% Benzene in petroleum ether. . . . . 10 ml
2. 50% Benzene in petroleum ether. . . . . 15 ml
3. 75% Benzene in petroleum ether. . . . . 20 ml
4. Benzene . . . . 20 ml

Estradiol diacetate should be eluted completely between the second and third 5 ml fraction of mixture No. 3. The acetates of estrone and estriol are eluted quantitatively by the first 15 ml of benzene. Analysis is carried out on GLC with a 3% SE-30 column which allows the complete separation of estrone and estradiol acetates. If the alumina is properly standardized, there should not be admixture of these two estrogens. Once a batch of alumina is standardized, it can be stored in an airtight container.

Gas Chromatography. Columns were made from commercially prepared coated material (QF-1, 3% on 80-100 mesh Gas Chrom P, Applied Science Laboratories, Inc.), and for SE-30, SE-30 and EGA, and NGS by filtration or evaporation of an alcohol-washed commercial 80-100 mesh diatomaceous earth (Gas Chrom Z, Applied Science Laboratories, Inc., and Diatoport S, F & M Scientific Corp.). Spiral columns are packed under vacuum with limited vibration. U-tubes are packed by gravity and limited vibration. SE-30 columns are

TABLE 18-15. Comparison of Reliability Criteria of Methods For Urinary Estrogen Determinations

Method	Estrogens	Accuracy	Precision	Sensitivity	Specificity	Speed
	Classical estrogens:					
Brown (1955)	Estrone Estradiol-17β Estriol	Recovery of 80-90% at 4 μg per 24 hr	± 15% or less with 5 μg per 24 hr	5 μg per 24 hr	Solvent partition, derivatives, adsorption chromatography, color reaction	(1 technician) 6 Urines per 6 days
Bauld (1956)	Estrone Estradiol-17β Estriol	Recovery of 80-90% at 4 μg per 24 hr	± 15% or less with 5 μg per 24 hr	5 μg per 24 hr	Solvent partition, column partition chromatography, color reaction	5 Urines per 6 days
Preedy and Aitken (1961)	Estrone Estradiol-17β Estriol	Recovery of 80-90% at 4 μg per 24 hr	Not available	5 μg per 24 hr	Solvent partition, column partition chromatography, fluorescence	Not available

Method	Compound	Recovery	Precision	Sensitivity	Methods	Urines
Method E-IV	Estrone Estradiol-17β Estriol	72.4–82.8% when 1 or 0.5 µg were added	±4.2% or less by replicate analysis after addition of known amount	0.2 µg per 24 hr	Solvent partition, derivatives, adsorption chromatography, gas—liquid chromatography	15 Urines per 5 days
Newer metabolites:						
Givner (1960)	2-Methoxyestrone Estrone 16α-Hydroxyestrone Estradiol-17β 16-Epiestriol Estriol	95% 90% 65% 66% 87% 76%	Not available	5 µg per 24 hr	Girard complex formation, solvent partition, column partition chromatography, color reaction	4 Urines per 8 days
Method E-III	2-Methoxyestrone Estrone 16α-Hydroxyestrone 16-Ketoestradiol-17β Estradiol-17β 16-Epiestriol Estriol	78% 83% 91% 91% 84% 72% 82%	±11% or less by replicate analysis after addition of 1 µg of each steroid	0.2 µg per 24 hr	Thin-layer chromatography, derivative formation, solvent partition, gas—liquid chromatography	12 Urines per 5 days

cured at 300°C for ～2 hr without gas flow, followed by 250°C with
10 psi N_2. All other columns are cured at 240°C, with 10 psi N_2 for
48 hr.

Injection of either 1 or 2 μl of the extracts is usually made after
saturation of the column with standards. Quantitative measurement
is carried out by comparison of the peak height of the unknown
standard at a concentration approximating the former.

Column conditions are quite variable and the reader is referred
to the individual figures describing the chromatograms.

Method E-I: The Rapid Determination of Estrone and Estriol in Pregnancy Urine (Wotiz and Martin)

For third trimester pregnancy urines, a 50 ml urine sample is
generally quite adequate, while specimens with much lower concentrations of estrogens may require larger amounts to be extracted.

Hydrolysis and Extraction. Although much evidence has been presented that acid hydrolysis sacrifices large quantities of estrogens
(50–60%), the vagaries of enzymatic hydrolysis often deter the investigator from using this less destructive procedure. The choice of
hydrolysis therefore rests with the individual investigator. For the
sake of conformity, the acid cleavage (as described by Brown) is
applied to a 50 ml urine aliquot after dilution with an equal volume
of water.

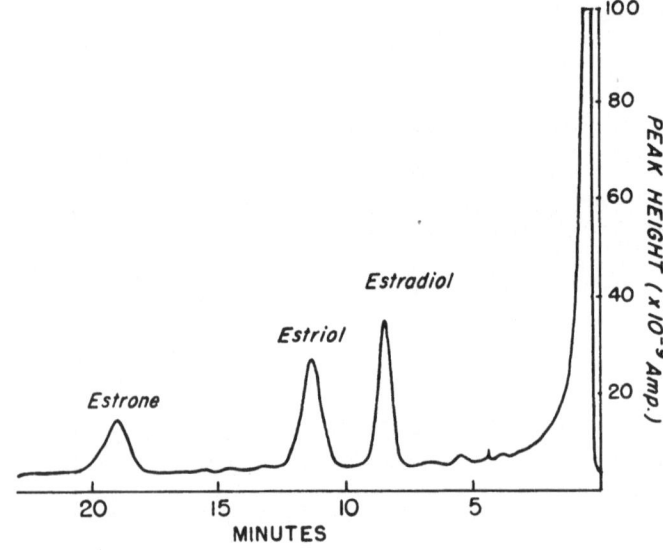

Figure 18-16. Gas chromatographic separation (on 3% XE-60) of estrogen heptafluorobutyrates
$(10^{-10}g)$ and measurement with an electron capture detector.

TABLE 18-16. Rapid Analysis of Estrone and Estriol in Pregnancy

1. Acid hydrolysis
2. Ether extraction
3. Na_2CO_3 wash [Brown (1955)]
4. Evaporation of ether
5. Partition between benzene–petroleum ether and 1 N NaOH
6. Re-extraction of aqueous solution at pH 9–10 with ether
7. Evaporation of ether
8. Acetylation with 0.1 ml pyridine and 0.5 ml acetic anhydride for 1 hr at 68°C
9. GLC on 6 ft by 4 mm 3% SE-30 column (3000 plates or better) at 250°C

The specimen is brought to a boil with 15 vol. % (7.5 ml) of concentrated HCl and the mixture refluxed for $\frac{1}{2}$ hr. After cooling in a stream of water, the urine is transferred to a separatory funnel and extracted three times with ether (1 × 100 ml, 2 × 50 ml). The combined ether extract is washed once with 50 ml of 8% $NaHCO_3$. The estrogens in the ether extract are partitioned into 2 × 50 ml of 1 N NaOH. The hydroxide layers are immediately neutralized with HCl and the phenolic fraction is re-extracted with three portions of ether as before. The ether extract is filtered through anhydrous Na_2SO_4 and then taken to dryness on a flash evaporator. With a little acetone the dried residue is transferred to a small (1 dram) screw-cap vial. This solution is again dried in a stream of N_2 or air.* To the dried residue is added 1 ml of dry acetic anhydride and 0.2 ml of dry pyridine, the mouth of the vial is covered with aluminum foil, and the screw cap is sealed tightly. The tube is immersed in a water bath (or air bath) at 60°–70°C for $\frac{1}{2}$ hr, after which time the contents of the tube are again evaporated to dryness in a stream of N_2 while being kept in a fume hood at about 60°C. To the dried residue is added 0.1 ml of acetone and the top is replaced. The sample is now ready for GLC.

Injection. Usually 1 or 2 μl is drawn into the capillary pipette or a Hamilton syringe. In case of the former, a small tissue (Wipette) is wetted with solvent and the outside of the pipette is wiped gently. Care must be taken not to touch the orifice or the sample will discharge. The amount to be injected depends largely on the concentration of the components to be analyzed. Other factors must, however, be taken into consideration. An excessively large injection may cause serious overlap of peaks and may obscure minor peaks. It may also prevent the detector from approaching baseline to give reading at a sufficiently low attenuation. It is generally more desirable to inject small volumes and decrease the attenuation.

Column Conditions. A 6 ft, 2–4 mm ID, 3% SE-30 on Gas Chrom P (80-100 mesh) column is generally used. The column temperature

*Note: Careful removal of all water at this point is essential to complete acetylation.

TABLE 18-17. Measurement of Classical Estrogens in Pregnancy or
Nonpregnancy Urine

1. Acid hydrolysis
2. Ether extraction
3. Na_2CO_3 wash [Brown (1955)]
4. Evaporate ether to dryness
5. Separation of phenolic and neutral fractions by partition between benzene—petroleum ether and 1 N NaOH
6. Re-extraction of the aqueous solution at pH 9.5—10 with ether
7. Evaporation
8. Acetylation with 0.1 ml pyridine and 0.5 ml acetic anhydride for 1 hr at 68°C
9. Solvent partition
10. Alumina column: (a) 25% benzene in petroleum ether, 10 ml—discarded
    (b) 50% benzene in petroleum ether, 15 ml—discarded
    (c) 75% benzene in petroleum ether, 20 ml: first 5 ml — discarded, next 10 ml — E_2A, last few ml — discarded
    (d) benzene, 20 ml: first 15 ml — $E_1A + E_3A$
11. GLC on 6 ft by 4 mm, 3% QF-1 column (3000 plates or better)

TABLE 18-18. Measurement of Classical Estrogens and Newer
Metabolites [40]

1. Enzyme hydrolysis
2. Ether extraction
3. Wash with 8% $NaHCO_3$, then water
4. Evaporate ether to dryness
5. TLC in System I [benzene : ethyl acetate (1 : 1) incompletely saturated]
6. Fraction A: estrone, 2-methoxyestrone
   Fraction B: estradiol, ring D-α-ketols (16α-hydroxyestrone, 16-ketoestradiol)
   Fraction C: 16-epiestriol
   Fraction D: estriol
7. Elution of individual fractions with ethanol
8. TLC in System II of Fraction B (remove interfering androgens)
9. Elution with ethanol
10. All fractions to dryness
11. Acetylation with 0.1 ml pyridine and 0.5 ml acetic anhydride for 1 hr at 68°C
12. Addition of 5 ml of water and extraction with petroleum ether
13. Evaporation of petroleum ether, redissolved in 50 or 100 μl of acetone
14. GLC 3% SE-30 on 80-100 mesh alcohol washed Diatoport S,\4 ft by 4 mm column. Fractions A and B : column temperature 228°C, 20 psi N_2; Fractions C and D : column temperature 238°C, 20 psi N_2

varies from 250° to 265°C, with the injector port about 20°C higher and the detector (flame) temperature at 300°C. A nitrogen flow rate of ~70 ml/min is usually maintained.

Measurement. Experiments have shown that within useful limits of the instrument the peak width remains constant and the area under the peak is therefore related to the peak height. Indeed, with some practice, one can achieve ±2% reproducibility of measurement based on peak height measurement only. It is best to run standards of known concentration through the instrument each morning and measure the peak heights. Repeat measurements at midday and evening will confirm the data. More accurate information may require more frequent measurement of standard or the incorporation of a known standard into every sample.

Computation of concentrations is of course the simple ratio:

$$\frac{C_s}{H_s} = \frac{C_u}{H_u}$$

where C_s is the concentration of standard, H_s is the peak height of standard, C_u is the concentration of unknown, and H_u is the peak height of unknown.

Method E-II: Determination of the Three Classical Estrogens Throughout Pregnancy Using Alumina and Gas-Liquid Chromatography (Chattoraj and Wotiz)

The details of this method are the same as those listed in Method E-IV with the exception that only $\frac{1}{20}$ of a 24 hr urine collection is used for analysis.

Method E-III: Determination of Seven Estrogens in Pregnancy Using Thin-Layer and Gas-Liquid Chromatography (Wotiz and Chattoraj)

Hydrolysis and Extraction. The method described by Givner et al. [20] was adopted. To an aliquot of urine, brought to pH 5.2 with 0.1 N acetic acid, is added 500 units of "Glusulase" per ml of urine, and the mixture incubated for 24 hr at 39 ± 0.5°. The urine is cooled to 5°C, diluted fourfold with distilled water, and transferred to a separatory funnel.

An extract is prepared by partitioning the lipids into diethyl ether, once with an equal volume and twice with half the volume of urine. The combined ether extract is then washed with 8% $NaHCO_3$ (1 × 10 ml per 100 ml) followed by water washes (2 × 5 ml per 100 ml) until the discard is neutral. After the extract is dried over Na_2SO_4, the ether is evaporated to dryness on a rotating still.

Thin-Layer Chromatography. The dried residue, dissolved in 100 μl of absolute ethanol, is applied as a thin streak on a previously prepared thin-layer plate along with two spots of a mixture of estrone, estradiol, estriol, and 16-epiestriol on both sides of the sample. The chromatogram is developed in System I (1:1, benzene:ethyl acetate, incompletely saturated). The area of the chromatogram taken up by the urine extract is covered with the template and the two edges containing the standard sprayed with Folin–Ciocalteau reagent. The extract is resolved into four different fractions:

Fraction 1. . . .Estrone and 2-methoxyestrone
Fraction 2. . . .Estradiol and ring-D-α-ketols
Fraction 3. . . .16-Epiestriol
Fraction 4. . . .Estriol

After elution with several portions of alcohol, the fraction containing estradiol and ring-D-α-ketols is further chromatographed in System II (10:9:1, petroleum ether: dichloromethane:ethanol, completely saturated). The zones corresponding to the standards are again scraped from the plate and eluted as before.

Acetylation and Solvent Partition. Each of the four zones eluted from the two TLC's is dried in a small vial under a stream of nitrogen and the residue incubated with a mixture of 5 parts of acetic anhydride (0.1 ml) and 1 part of pyridine (0.02 ml) at 68°C for 1 hr. To the acetylated mixture 10 ml of distilled water is added, while stirring thoroughly with a glass rod. The sample is then transferred to a separatory funnel and extracted once with 10 ml and twice with 5 ml of light petroleum ether. The vessel used for acetylation is rinsed with petroleum ether after each transfer and the rinsings added to the separatory funnel. The combined petroleum ether is washed once with 8% $NaHCO_3$ (5 ml) followed by 3 ml portions of water until the discard is neutral. The solvent is evaporated to dryness, the residue transferred with petroleum ether or dichloromethane to a 2 ml centrifuge tube and evaporated once more.

Gas Chromatography. Each residue is dissolved in the appropriate amount of n-hexane (50 μl) and an aliquot of this solution (2–5 μl) is gas chromatographed on a 3% SE-30 column (4 ft by $\frac{1}{8}$ in.) at 228° or 238°C. If the 3% SE-30 column is used, it will not be possible to separate 16 α-hydroxyestrone from 16-ketoestradiol. Should this separation be desired, then the fraction from the thin-layer chromatogram containing the ring-D-α-ketols along with estradiol should be chromatographed on a 4-ft, $\frac{1}{8}$-in. column using 80-100 mesh support (Gas Chrom Q or Diatoport S) coated first with 0.5% EGA (stabilized), followed by a second coating with 2.5% SE-30. Separation of the two isomers can be obtained on such a column.

Quantification by comparison of peak heights from the chromatogram of the extract with the corresponding standard is carried out as described in the section "Rapid Determination of Estrone and Estriol in Crude Extracts of Late Pregnancy Urine."*

Method E-IV: Analysis of Estrone, Estradiol, and Estriol in Low-Titer Urines (Chattoraj and Wotiz)

One-fifth of the 24 hr collection of urine is diluted with an equal volume of distilled water. Acid hydrolysis, extraction, and separation into the phenolic and nonphenolic fractions are carried out according to the method of Brown [2], the only exception being the use of 1 N NaOH for the extraction of three classic estrogens in the same fraction (see Method E-I). After the pH of the alkaline solution has been altered to between 9.5 and 10 by the addition of solid $NaHCO_3$, the mixture is extracted, once with an equal volume of ether and twice with half the volume. The combined ether layers are washed with 8% $NaHCO_3$ solution (5 ml per 100 ml) followed by water (2.5 ml per 100 ml) until the discard is neutral. Following evaporation to dryness in a rotating still under reduced pressure, the dried phenolic residue is transferred to a test tube with a little dichloromethane.

Acetylation. The residue is acetylated by dissolving it in 1 ml of a mixture of 5 parts of acetic anhydride and 1 part of pyridine and keeping the tube at 68°C for 1 hr. To the acetylated mixture 10 ml of distilled water is added, while stirring thoroughly with a glass rod. The sample is then transferred to a small separatory funnel and extracted twice with 10 ml and once with 5 ml of light petroleum ether. The vessel used for acetylation is also rinsed with petroleum ether and the washings added to the separatory funnel. The combined petroleum ether fractions are washed with 8% $NaHCO_3$ (5 ml) followed by 2 ml portions of water until the discard is neutral. The solution is allowed to stand for 10 min and any water collected at the bottom is removed as completely as possible.

Alumina Column Chromatography. The preparation of the column and the elution of the estrogens are followed exactly as described in the section "Materials and Procedures."† The proper eluates containing the respective estrogens are evaporated to dryness in a 50 ml round bottom flask on a rotating still. The residues are transferred to 2 ml centrifuge tubes with petroleum ether or hexane. The solvent is dried and the residue is redissolved in 20 μl of hexane.

Then, 2 μl of this solution is injected into a 6-ft column packed with 3% QF-1 on 80-100 mesh support and measurements taken by comparison to standards as described previously.

*See page 261.
†See page 255.

Tables 18-16, 18-17, and 18-18 are included as outlines of the three different procedures so that the reader may more readily examine the complexity of each individual assay to assist him in his choice.

REFERENCES

1. H. Breuer, Vitamins Hormones 20:307 (1962).
2. J. B. Brown, Biochem. J. 60:185 (1955).
3. W. S. Bauld, Biochem. J. 68:488 (1956).
4. G. Ittrich, Z. Physiol. Chem. 312:1 (1958).
5. J. R. K. Preedy, Estrogens, in: Methods in Hormone Research, edited by R. I. Dorfman (Academic Press, New York and London, 1962).
6. W. S. Bauld and Greenway, in: Methods of Biochemical Analysis, edited by D. Glick (Interscience Publishers, New York, 1957), Vol. V, p. 337.
7. W. J. A. VandenHeuvel, C. C. Sweeley, and E. C. Horning, J. Am. Chem. Soc. 82:3481 (1960).
8. H. H. Wotiz and H. F. Martin, 138th Am. Chem. Soc. Meeting, Abstract p. 58C, New York (1960).
9. H. H. Wotiz and H. F. Martin, Anal. Biochem. 3:97 (1962).
10. J. Fishman and J. B. Brown, J. Chromatog. 8:21 (1962).
11. T. Luukkainen, W. J. A. VandenHeuvel, E. O. A. Haahti, and E. C. Horning, Biochim. Biophys. Acta 52:599 (1961).
12. J. L. Yousem, Am. J. Obstet. Gynecol. 88:375 (1964).
13. R. I. Cox and A. R. Bedford, Steroids 3:663 (1964).
14. J. C. Touchstone, H. H. Veron, and T. Murawec, Biochemistry 3:126 (1964).
15. J. C. Touchstone, J. Gas Chromatog. 2:170 (1964).
16. W. J. A. VandenHeuvel, J. Sjövall, and E. C. Horning, Biochim. Biophys. Acta 48:596 (1961).
17. H. H. Wotiz and H. F. Martin, Federation Proc. 20:199 (1961).
18. H. H. Wotiz and S. C. Chattoraj, Anal. Chem. 36:1466 (1964).
19. S. J. Clark and H. H. Wotiz, Steroids 2:540 (1963).
20. M. L. Givner, W. S. Bauld, and W. Kitty, Biochem. J. 77:406 (1960).

Chapter 19

Hopes and Needs for the Future

Interest in the estimation of steroids by gas chromatographic methods continues to grow, judging from the number of papers appearing in the literature. It seems appropriate to end this book with a very brief summary of the achievements and failures of the technique and to indicate what is likely to be done in the immediate future.

From the purely gas chromatographic point of view, much has been achieved in the past few years. Instrument performance and reliability have improved and manufacturers in general provide adequate service. In addition, the better manufacturers maintain extensive applications laboratories which can provide considerable assistance in the solution of specific problems. Major advances have been made in column technology, particularly in the production of inert support material. The combination of improved instrumentation and columns has resulted in a doubling in the column efficiency that can be achieved routinely. At least one manufacturer now offers columns guaranteed to have efficiencies of at least 500 plates per foot, where two or three years ago columns of this quality for steroid analysis were obtained largely by chance.

The technology appears to have entered a period of stability and it is to be expected that steady refinement and consolidation rather than startling advances will be the pattern for some time to come.

A general problem in gas chromatography is standardization of methods so that interlaboratory comparison of results is possible and meaningful. Some progress has been made toward standardization of retention ratios for simple compounds separated on stationary phases that are single chemical compounds. The use of polymeric phases at low percentages, as in steroid gas chromatography, causes difficulties both because of the lack of chemical identity of the stationary liquid and because of activity of the support. It is hoped that work on the characterization of column materials will continue so that retention time data become more reliable for the qualitative analysis of steroid molecules. It must be realized that, without special precautions, gas chromatography is less precise than many other analytical techniques.

265

However, the precision attainable with a relatively simple apparatus is more than adequate for biochemical work.

A perusal of the methods described will show that the assay of many steroid hormones is now possible on a routine basis. For the more stable compounds at high levels, the procedures are simple and can be entrusted to a relatively inexperienced technician. The techniques lend themselves well to automation; it is expected that the simple procedures will soon be automated, with the possible exception of the hydrolysis step. When this has been accomplished, the methods will take their place alongside other automatic procedures now in routine use.

Methods for low-level determination both in urine and plasma are more difficult and at present require considerable skill if valid results are to be obtained. At the same time it must be emphasized that no alternative exists to some of these methods and thus these represent a new tool for the research worker and clinician. For some compounds, no entirely satisfactory method as yet exists. The difficulty of chromatographing intact the glucocorticoids has so far limited the applicability of gas chromatography to the estimation of these compounds. For the same reason difficulties have been experienced in the determination of tetrahydroaldosterone in urine. Furthermore, the conditions for the preliminary paper chromatographic separation of this metabolite are critical. No method as yet exists for the estimation of aldosterone; the instability of this compound and the low levels sought make development of a method extremely difficult. Other compounds for which methods are needed will doubtlessly occur to the reader. Because of the considerable activity in the field, it is to be expected that many new methods will become available during the next few years.

One property of the gas chromatographic system that may be emphasized is its relative nonspecificity. A high-resolution column coupled to a nonselective flame ionization detector provides far more information than is used in the estimation of a single compound. The study of chromatographic records for hitherto unknown or unsuspected correlations could be profitable and could lead to the development of a method in reverse, where quantitative values are available before the identity of a compound is confirmed. It is up to the alert investigator to take advantage of such chance.

Directory of Suppliers

The directory lists the major suppliers of instruments and accessories. A more complete list can be found in J. Gas Chromatog., June 1965. Suppliers are generally listed under one heading only although many of the large instrument manufacturers supply most of the listed accessories and supplies.

INSTRUMENTS

AMERICAN INSTRUMENT CO., INC.
8030 Georgia Avenue
Silver Spring, Maryland 20910

BARBER-COLMAN COMPANY
Rockford, Illinois

BECKMAN INSTRUMENTS, INC.
2500 Harbor Boulevard
Fullerton, California 92634

BURRELL CORPORATION
2223 Fifth Avenue
Pittsburgh 19, Pennsylvania

CARY INSTRUMENTS
Applied Physics Corporation
2724 South Peck Road
Monrovia, California

F & M SCIENTIFIC CORP.
Route 41 and Starr Road
Avondale, Pennsylvania

GLOWALL CORPORATION
2530 Wyandotte Road
Willow Grove, Pennsylvania

JARRELL-ASH COMPANY
590 Lincoln Street
Waltham, Massachusetts 02154

LOENCO, INC.
2092 North Lincoln Avenue
Altadena, California 91002

MICRO-TEK INSTRUMENTS, INC.
P. O. Box 15409
Baton Rouge, Louisiana

NESTER/FAUST
2401 Ogletown Road
Newark, Delaware

PACKARD INSTRUMENT CO., INC.
2200 Warrenville Road
Downers Grove, Illinois 60515

THE PERKIN-ELMER CORP.
870 Main Avenue
Norwalk, Connecticut

WARNER-CHILCOTT LABORATORIES
Instrument Division
200 South Garrard Boulevard
Richmond, California

WILKENS INSTRUMENT & RESEARCH INC.
2700 Mitchell Drive, Box 313
Walnut Creek, California

BECKER-DELFT (Julian H. Becker N.V.)
Vulcanusweg 113, P. O. Box 219
Delft, Holland

BODENSEEWERK PERKIN ELMER & CO.
GmbH
777 Uberlingen/Bodensee
West Germany

CARLO ERBA S.p.A.
Via Carlo Imbonati, 24
Milano, Italy

GAS CHROMATOGRAPHY LIMITED
Boyn Valley Road, Maidenhead
Berkshire, England

W. G. PYE & CO. LTD.
P. O. Box 60, York Street
Cambridge, England

SCIENTIFIC KIT CO.
P. O. Box 244
Washington, Pennsylvania

SHANDON SCIENTIFIC COMPANY, LTD.
6 Cromwell Place
London SW 7, England

SIEMENS & HALSKE AKTIENGESELL-
SCHAFT
Wernerwerk fur Messtechnik
Rheinbruckenstrasse 50, P.O.B. 4480
75 Karlsruhe, West Germany

DR. VIRUS KG
Luftelberger Strasse
Meckenheim bei Bonn, Germany

COLUMN SUPPORT MATERIALS
AND STATIONARY PHASES

ANALABS, INCORPORATED
P. O. Box 5215
Hamden, Connecticut 06518

APPLIED SCIENCE LABORATORIES, INC.
P. O. Box 140
140 North Barnard Street
State College, Pennsylvania

THE BRITISH DRUG HOUSES LTD.
THE EALING CORPORATION
225 Massachusetts Avenue
Cambridge, Massachusetts 02140

COAST ENGINEERING LABORATORY
3755 Inglewood Avenue
Redondo Beach, California

W. H. CURTIN & COMPANY
1800 Sidney Street
Houston 1, Texas

THE EAGLE-PICHER COMPANY
American Building
Cincinnati 1, Ohio

JOHNS-MANVILLE PRODUCTS CORP.
Celite Division
22 East 40th Street
New York, New York

MAY & BAKER LIMITED
Dagenham Essex, England

MICROBEADS DIVISION
Cataphote Corporation
P. O. Box 2369
Jackson, Mississippi

SCHLESINGER CHEMICAL CORP.
580 Mineola Avenue
Carle Place, L.I., New York

DETECTORS

CARLE INSTRUMENTS, INC.
532 South Rose Street
Anaheim, California 92805

GOW-MAC INSTRUMENT COMPANY
100 Kings Road
Madison, New Jersey 07940
(Thermal conductivity, gas density)

IONICS RESEARCH, INC.
22 Sandalwood Drive
Houston, Texas 77024
(Ionization detectors)

PACKARD INSTRUMENT CO., INC.
2200 Warrenville Road
Downers Grove, Illinois 60515
(Radioisotope detectors)

RECORDERS

THE BRISTOL COMPANY
Waterbury 20, Connecticut

ESTERLINE ANGUS INSTRUMENT CO.,
INC.
Box 596
Indianapolis, Indiana 46206

HONEYWELL, INCORPORATED
Wayne and Windrim Avenues
Philadelphia 44, Pennsylvania
(Honeywell Controls, Ltd., England)

GEORGE KENT LIMITED
Luton, Bedfordshire
England

LEEDS & NORTHRUP CO.
4901 Stenton Ave.
Philadelphia 44, Pennsylvania

E. H. SARGENT & COMPANY
4647 West Foster Avenue
Chicago 30, Illinois

TEXAS INSTRUMENTS, INC.
3609 Buffalo Speedway
P. O. Box 66027
Houston 6, Texas

VARIAN ASSOCIATES
Recorder Division
611 Hansen Way
Palo Alto, California 94303

WESTRONICS, INCORPORATED
3605 McCart
Fort Worth, Texas

ELECTRONIC INTEGRATORS

DATEX CORPORATION
1307 South Myrtle Avenue
Monrovia, California

INFOTRONICS CORPORATION
7800 Westglen Drive
Houston, Texas 77042

SYRINGES

HAMILTON COMPANY
P. O. Box 307
Whittier, California

INFRARED ACCESSORIES

BARNES ENGINEERING COMPANY
Instrument Division
Stamford, Connecticut

WILKS SCIENTIFIC CORPORATION
140 Water Street
South Norwalk, Connecticut

PIPE AND TUBE FITTINGS

CAJON COMPANY
3550 Old South Miles Road
Solon 39, Ohio

CRAWFORD FITTING COMPANY
884 East 140th Street
Cleveland 10, Ohio

HOKE INCORPORATED
1 Tenakill Park
Cresskill, New Jersey

PARKER-HANNIFIN CORPORATION
17325 Euclid Avenue
Cleveland, Ohio 44112

COLUMN TUBING

SUPERIOR TUBE COMPANY
Norristown, Pennsylvania

GASES

AIR PRODUCTS AND CHEMICALS, INC.
Allentown, Pennsylvania

AIR REDUCTION SALES COMPANY
Div. of Air Reduction Co., Inc.
150 East 42nd Street
New York 17, New York

LINDE AIR PRODUCTS COMPANY
A Division of Union Carbide Corp.
East Park Drive and Woodward Avenue
Tonawanda, New York

THE MATHESON COMPANY, INC.
P. O. Box 85
East Rutherford, New Jersey

PRECISION GAS PRODUCTS, INC.
Box 364
Westfield, New Jersey 07091

Bibliography

BASIC TEXT BOOKS

GAS CHROMATOGRAPHY, D. Ambrose and Barbara A. Ambrose, 220 pp, $6.75 (D. Van Nostrand Co., Inc., 1962).

GAS CHROMATOGRAPHY, Ernst Bayer, 240 pp, $5.00 (English translation of 1st German ed.) (American Elsevier Publishing Co., Inc., 1961).

GAS PHASE CHROMATOGRAPHY, 3 Vols. (in English), Rudolph Kaiser, translated by P. H. Scott. Vol. I, GAS CHROMATOGRAPHY, 199 pp, $7.95, Vol. II, CAPILLARY CHROMATOGRAPHY, 120 pp, $6.95, Vol. III, TABLES FOR GAS CHROMATOGRAPHY, 162 pp, $7.75 (Butterworths, Inc., 1963).

GAS CHROMATOGRAPHY, A. I. M. Keulemans, 217 pp (2nd ed. 1959), 234 pp, $7.50 (Reinhold Publishing Corp., 1957).

GAS CHROMATOGRAPHY, J. H. Knox, 126 pp, $3.25 (John Wiley & Sons, Inc., 1962).

GAS CHROMATOGRAPHY: PRINCIPLES, TECHNIQUES AND APPLICATIONS, A. B. Littlewood, 507 pp, $15.00 (Academic Press, Inc., 1962).

GAS LIQUID CHROMATOGRAPHY: THEORY AND PRACTICE, Stephen Dal Nogare and Richard S. Juvet, 450 pp, $13.95 (John Wiley & Sons, Inc., 1962).

GAS CHROMATOGRAPHY, Howard Purnell, 441 pp, $12.00 (John Wiley & Sons, Inc., 1962).

SYMPOSIA PROCEEDINGS

VAPOR PHASE CHROMATOGRAPHY — 1956, D. H. Desty, ed., 436 pp, $12.00.

GAS CHROMATOGRAPHY — 1958, D. H. Desty, ed., 383 pp, $13.00.

GAS CHROMATOGRAPHY — 1960, R. P. W. Scott, ed., 466 pp, $17.50 (Butterworths, Inc., 1960).

GAS CHROMATOGRAPHY — 1962, M. van Swaay, ed., 411 pp, $19.95 (Butterworths, Inc., 1963).

GAS CHROMATOGRAPHY: First International Symposium Held under the Auspices of the Analysis Instrumentation Division of the Instrument Society of America, Vicent J. Coates, Henry J. Noebels, and Irving S. Fagerson, eds., 328 pp, $12.00 (Academic Press, Inc., 1958).

GAS CHROMATOGRAPHY: Second International Symposium Held under the Auspices of the Instrument Society of America, Henry Noebels, R. F. Wall, and Nathaniel Brenner, eds., 463 pp, $16.00 (Academic Press, Inc., 1961).

GAS CHROMATOGRAPHY: Third International Symposium Held under the Auspices of the Analysis Instrumentation Division of the Instrument Society of America, Nathaniel Brenner, Joseph E. Callan, and Marvin D. Weiss, eds., 719 pp, $22.00 (Academic Press, Inc., 1962).

271

GAS CHROMATOGRAPHY: Fourth International Symposium, Analysis Instrumentation Division, Instrument Society of America, June 1963, Lewis Fowler, ed., 270 pp, $10.50 (Academic Press, Inc., 1963).

APPLICATIONS

BIOCHEMICAL APPLICATIONS OF GAS CHROMATOGRAPHY, H. P. Burchfield and Eleanor E. Storrs, 680 pp, $22.00 (Academic Press, Inc., 1962).

STEROID CHROMATOGRAPHY, Robert Neher, approx. 275 pp, $11.00 (American Elsevier Publishing Co., Inc., 1964).

BIOMEDICAL APPLICATIONS OF GAS CHROMATOGRAPHY, H. A. Szymanski, ed., 330 pp, $12.50 (Plenum Press, 1964).

LECTURES ON GAS CHROMATOGRAPHY — 1964: AGRICULTURAL AND BIOLOGICAL APPLICATIONS, H. A. Szymanski and L. R. Mattick, eds., 300 pp, $12.50 (Plenum Press, 1965).

CHROMATOGRAPHY, Erich Heftmann, ed., 753 pp, $17.50 (Reinhold Publishing Corp., 1965).

OPEN TUBULAR COLUMNS IN GAS CHROMATOGRAPHY, L. S. Ettre, 184 pp, $4.95 (Plenum Press, 1965).

REVIEWS

CHROMATOGRAPHY: A REVIEW OF PRINCIPLES AND APPLICATIONS, E. Lederer and M. Lederer, 2nd ed. (1st ed. 1955), 712 pp, $14.00 (American Elsevier Publishing Co., Inc., 1957).

CHROMATOGRAPHIC REVIEWS, M. Lederer (American Elsevier Publishing Co., Inc.). Vol. II, 195 pp, $9.00 (1959), Vol. IV, 184 pp, $9.50 (1961), Vol. V, 244 pp, $11.00 (1962), Vol. VI, 228 pp, $12.50 (1963).

JOURNALS

JOURNAL OF CHROMATOGRAPHY, Vols. 1—4, per vol. $18.00 (1958—1960), Vols. 5—6, per vol. $18.00 (1961), Vols. 7—9, per vol. $18.00 (1962), Vols. 10—12, per vol. $15.00 (1963), Vols. 13—16, per vol. $15.00 (1964). Single issues, per 100 pp, $2.80. Vols. 17—20, per vol. $17.50 (1965) (American Elsevier Publishing Co., Inc.).

JOURNAL OF GAS CHROMATOGRAPHY. Published monthly. One year subscription: $10.00 in the U. S.; $12.00 all other countries (Preston Technical Abstract Co.).

ABSTRACTS

GAS CHROMATOGRAPHY ABSTRACTS — 1958, C. E. H. Knapman, ed., 262 pp, $8.50 (1960).

GAS CHROMATOGRAPHY ABSTRACTS — 1959, C. E. H. Knapman, ed., 164 pp, $8.50 (1960).

GAS CHROMATOGRAPHY ABSTRACTS — 1960, C. E. H. Knapman, ed., 164 pp, $8.50 (1960).

GAS CHROMATOGRAPHY ABSTRACTS — 1961, C. E. H. Knapman, ed., 219 pp, $8.50 (1962).

GAS CHROMATOGRAPHY ABSTRACTS — 1962, C. E. H. Knapman, ed., 203 pp, $8.50 (1963) (Butterworths, Inc.).

GAS CHROMATOGRAPHY ABSTRACT SERVICE. Published weekly on punched cards (Preston Technical Abstract Co.).

COMPREHENSIVE BIBLIOGRAPHY TO THE LITERATURE ON GAS CHROMATOGRAPHY, Seaton T. Preston, Jr., and Geneva Hyder, $12.00 (Preston Technical Abstract Co., 1964).

GUIDE TO GAS CHROMATOGRAPHY LITERATURE, A. V. Signeur, 360 pp, $12.50 (Plenum Press, 1964).

GUIDE TO GAS CHROMATOGRAPHY LITERATURE. SUPPLEMENT NO. 1, A. V. Signeur, approx. 350 pp, about $12.50 (Plenum Press, 1965).

Author Index

275

Subject Index

279